Student Resource Guide

DALE R. BUSKE

St. Cloud State University

Excursions IN MODERN MATHEMATICS

Sixth Edition

PETER TANNENBAUM

PEARSON

Prentice Hall

Upper Saddle River, NJ 07458

Editor-in-Chief: Sally Yagan
Acquisitions Editor: Chuck Synovec
Supplement Editor: Joanne Wendelken
Executive Managing Editor: Kathleen Schiaparelli
Senior Managing Editor: Nicole M. Jackson
Assistant Managing Editor: Karen Bosch Petrov
Production Editor: Dana Dunn
Supplement Cover Manager: Paul Gourhan
Supplement Cover Designer: Christopher Kossa
Manufacturing Buyer: Ilene Kahn
Manufacturing Manager: Alexis Heydt-Long

© 2007 Pearson Education, Inc.
Pearson Prentice Hall
Pearson Education, Inc.
Upper Saddle River, NJ 07458

Pearson Prentice Hall™ is a trademark of Pearson Education, Inc.

The author and publisher of this book have used their best efforts in preparing this book. These efforts include the development, research, and testing of the theories and programs to determine their effectiveness. The author and publisher make no warranty of any kind, expressed or implied, with regard to these programs or the documentation contained in this book. The author and publisher shall not be liable in any event for incidental or consequential damages in connection with, or arising out of, the furnishing, performance, or use of these programs.

Printed in the United States of America

10   9   8   7   6   5   4   3   2   1

ISBN   0-13-187382-2

Pearson Education Ltd., *London*
Pearson Education Australia Pty. Ltd., *Sydney*
Pearson Education Singapore, Pte. Ltd.
Pearson Education North Asia Ltd., *Hong Kong*
Pearson Education Canada, Inc., *Toronto*
Pearson Educación de Mexico, S.A. de C.V.
Pearson Education—Japan, *Tokyo*
Pearson Education Malaysia, Pte. Ltd.

Table of Contents

Apportionment Today

The Huntington-Hill Method

Chapter 4 made for a rugged but illuminating mathematical and historical excursion. But no apportionment excursion would be complete (at least from an American perspective) without an additional trip into the present. The purpose of this mini-excursion is to discuss the modern method of apportionment of the U.S. House of Representatives. In so doing, we will venture into some new mathematical territory.

Historical Background

By 1929 Congress was fed up with the controversies surrounding the decennial apportionment of the House of Representatives, and there was a strong push within Congress to place the issue on a firm mathematical foundation. At the formal request of the Speaker of the House of Representatives, Nicholas Longworth, the National Academy of Sciences commissioned a panel of distinguished mathematicians to investigate the different apportionment methods that had been proposed over the years and to recommend, in the words of Representative Gibson, the "correct mathematical formula." After deliberate consideration and discussion, the panel endorsed a method then called the **method of equal proportions**, a method originally proposed sometime around 1911 by Joseph A. Hill, the Chief Statistician of the Bureau of Census, and later improved and refined by Edward V. Huntington, a Professor of Mechanics and Mathematics at Harvard University. The method is now known as the **Huntington-Hill method** (or some variation of this name, such as the *Hill method* or the *Hill-Huntington method*).

It took a while, but eventually the politicians paid heed to the mathematicians. In 1941, Congress passed, and President Franklin D. Roosevelt signed, "An Act to Provide for Apportioning Representatives in Congress among the Several States by the Equal Proportions Method," generally known as *the 1941 apportionment act*. This act had three key elements: (i) It set the Huntington-Hill method as the *permanent* method for apportionment of the House of Representatives, (ii) it made the decennial apportionments *self-executing* (once the official population figures are in, the apportionment formula is applied automatically and changes take effect without the need for Congressional approval), and (iii) it permanently fixed the size of the House of Representatives at 435 seats (with an exception made if a new state were to join the Union).

With a "fixed-size" House, one state's gain must be another state's loss, and, typically, the states losing those seats tend to put up a fight. These fights are fought in the federal courts (including the Supreme Court—see the Project at the end of this mini-excursion), and ultimately the legal arguments always boil down to a core mathematical question—what is the "correct mathematical formula" for apportionment? States losing seats under Huntington-Hill always try to make the argument that the Huntington-Hill method is not it, and that some other method (whichever one lets them keep their seat) is the correct mathematical formula. Ultimately, the courts have ruled against these challenges, and the Huntington-Hill method remains the method of choice—at least until the next lawsuit.

Before we embark on a full-fledged description of the Huntington-Hill method, we will take a brief mathematical detour.

The Arithmetic and Geometric Means

The standard way of "averaging" two numbers (we are taught sometime in elementary school) is to take the number halfway between them. This kind of average is called the **arithmetic mean** of the two numbers. Less known or understood is the fact that there are other important and useful ways to "average" numbers. To avoid confusion with the conventional use of the word *average*, these other type of "averages" are referred to as **means**. Examples of important means beyond the arithmetic mean are the *geometric mean*, the *harmonic mean*, and the *quadratic mean*. Of all of these, the *geometric mean* is of particular relevance to our discussion.

See Exercises 21 and 22.

Geometric Mean

The *geometric mean* of two positive numbers a and b is the number $G = \sqrt{a \cdot b}$.

The requirement that the numbers be positive ensures that we don't have to deal with negative square roots.

We are going to do a lot of comparing between the geometric and arithmetic means of the same numbers, so for good measure we are including a formal definition of the arithmetic mean as well.

Arithmetic Mean

The *arithmetic mean* of two numbers a and b is the number $A = \dfrac{a + b}{2}$.

Here there is no requirement that the numbers be positive, but in our examples and applications they will turn out to be so.

One useful way to think about the difference between the arithmetic and geometric means is that the arithmetic mean is an average under *addition*, while the geometric mean is an average under *multiplication*.

▶ EXAMPLE 1.1 Averaging Growth Rates

Over the last two years, the student body at Tasmania State University grew at a rate of 5% during the first year and 75% during the second year. What was the *average annual growth rate* over the two-year period?

Tempting as it may seem to say that the average annual growth rate was 40% $[(5\% + 75\%)/2]$, the correct average growth rate was roughly 35.55%. To explain where this number came from we will work our way backwards: we will first look at the calculation, and then give a brief explanation of why we do it this way.

To compute the average growth rate, we first find the *geometric mean* of 1.05 and 1.75. This is given by $\sqrt{(1.05)(1.75)} \approx 1.35554$. From this number we deduce that the average annual growth rate is approximately 35.55%. Let's first deal with the 1's that mysteriously showed up. The 1.05 and the 1.75 are *multipliers* corresponding to the *growth rates* of 5% and 75%, respectively. [A growth rate of 5% means that in the first year a student population of P students grew to $(1.05)P$ students. A growth rate of 75% in the second year means that a student population of $(1.05)P$ students grew to $(1.75)(1.05)P$ students.] Likewise, the 1.35554 represents a growth rate of approximately 35.55%.

But why did we take the geometric mean of 1.05 and 1.75? The answer is that growth rates are not added, but rather compounded multiplicatively, and *when quantities are multiplied, the correct way to average them is to use the geometric mean*. In this case, what we really mean by the average growth rate is, *if the growth rate had been the same both years, what would it have been?* The answer is 35.55% because $(1.3555)(1.3555) \approx (1.05)(1.75)$.

The next pair of examples will highlight some important numerical differences between the arithmetic and geometric means.

EXAMPLE 1.2 The Arithmetic-Geometric Mean Inequality

Table 1-1 shows geometric means in the third column (rounded to decimal places when necessary) and arithmetic means in the last column. These values can be easily verified as long as you have a calculator handy (which you should).

What is Table 1-1 telling us? Think of a and b as starting as the same number (10 and 10). In this case the arithmetic and geometric means are equal ($G = A = a$). Now think of a staying put but b separating away from a and steadily getting bigger. As b moves away from a, both the geometric mean G and the arithmetic mean A fall somewhere in between a and b, but while A is always exactly halfway between a and b, G is closer to a than to b. (Think of G and A as two runners moving away from a, with A moving at a steady rate but G moving at a slower and slower rate as time goes on.)

See Exercises 10 and 19.

TABLE 1-1 Geometric Means Versus Arithmetic Means			
a	b	$G = \sqrt{a \cdot b}$	$A = (a + b)/2$
10	10	10	10
10	20	14.14	15
10	40	20	25
10	100	31.62	55
10	1000	100	505
10	9000	300	4505

The most important implication of our observation in Example 1.2 is that for any two positive numbers a and b, *the arithmetic mean is bigger than the geometric mean* except when $a = b$, in which case the two means are equal. This fact is known as the *arithmetic-geometric mean inequality*.

There are many ways to prove the arithmetic-geometric mean inequality —see Exercise 10. For a nice geometric proof using the Pythagorean Theorem, see Exercise 11.

Arithmetic-Geometric Mean Inequality

Let a and b be any two positive numbers (assume $a \leq b$) and let A and G denote their arithmetic and geometric means respectively. Then,

$$a \leq G \leq A \leq b.$$

(In the special case when $a = b$, we end up with $G = A$.)

EXAMPLE 1.3 Scaling Geometric and Arithmetic Means

Table 1-2 illustrates another simple but useful property of the geometric and arithmetic means: when a and b are multiplied by a common positive scaling factor k, their geometric and arithmetic means are also multiplied by the same scaling factor k. This fact can be proved using basic high school algebra.

See Exercise 9(a).

TABLE 1-2

a	b	*Geometric Mean*	*Arithmetic Mean*
1	4	2	2.5
10	40	20	25
0.25	1	0.5	0.625
1.5	6	3	3.75
k	$4k$	$2k$	$2.5k$

EXAMPLE 1.4 Geometric Means of Consecutive Integers

In this example we will examine the behavior of the geometric mean when the numbers a and b are consecutive integers. The geometric means here are rounded to three decimal places.

The last column of Table 1-3 shows the difference between the arithmetic and the geometric means. It's hard to miss the gist of the message of Table 1-3: *The difference between the arithmetic and the geometric means of two consecutive integers is very small, and as the integers get larger, this difference becomes even smaller.*

See Exercise 18.

TABLE 1-3			
a, b	*G*	*A*	*A − G*
1, 2	1.414	1.5	0.086
2, 3	2.449	2.5	0.051
3, 4	3.464	3.5	0.036
4, 5	4.472	4.5	0.028
5, 6	5.477	5.5	0.023
6, 7	6.481	6.5	0.019
7, 8	7.483	7.5	0.017
8, 9	8.485	8.5	0.015
9, 10	9.487	9.5	0.013
20, 21	20.494	20.5	0.006
100, 101	100.499	100.5	0.001

This is the end of our mathematical detour into the geometric mean, but obviously not the end of our dealings with the concept.

The Huntington-Hill Method

There are several ways to describe the Huntington-Hill method, but by far the easiest way to do it is by analogy to Webster's method. In fact, Huntington-Hill's method is a very close cousin of Webster's method, and we can say that the difference between the two methods boils down to *the difference between the arithmetic mean and the geometric mean.* We will start our discussion by briefly revisiting Webster's method.

Webster's Method Revisited

Without going into all the details (for details, see Section 4.6), the defining philosophy of Webster's method is that the state's *quotas are to be rounded to the nearest integer* (down if the decimal part is less than 0.5, up otherwise). The term *quota* here refers to either the *standard quota* (if that works then so much the better) or some *modified quota.*

A different way to describe the rounding rule for Webster's method rule is in terms of *cutoff points*—the cutoff point for rounding a quota is the midpoint (i.e., the *arithmetic mean*) between the *lower* and *upper* quotas—if the quota is less than this cutoff point then we round it down, otherwise we round it up.

The following formal description of Webster's method may seem like a bit of overkill, but it makes for a good precedent.

Webster's Rounding Rule

For a quota q, let L denote its lower quota, U its upper quota, and A the *arithmetic mean* of L and U. If $q < A$, then round q *down to L,* otherwise *round q up to U.*

Webster's Method

- **Step 1.** Find a "suitable" divisor D. [Here a suitable divisor means a divisor that produces an apportionment of exactly M seats when the *quotas* (populations divided by D) are rounded using *Webster's rounding rule*.]
- **Step 2.** Find the apportionment of each state by rounding its quota using Webster's rounding rule.

The description of the Huntington-Hill method follows exactly the same script but uses the *geometric mean* rather than the arithmetic mean to define the cutoff points.

Huntington-Hill Rounding Rule

For a quota q, let L denote its lower quota, U its upper quota, and G the *geometric mean* of L and U. If $q < G$, then round q *down to L*, otherwise *round q up to U*.

The Huntington-Hill Method

- **Step 1.** Find a "suitable" divisor D. [Here a suitable divisor means a divisor that produces an apportionment of exactly M seats when the *quotas* (populations divided by D) are rounded using the *Huntington-Hill rounding rule*.]
- **Step 2.** Find the apportionment of each state by rounding its quota using the Huntington-Hill rounding rule.

When doing apportionments under the Huntington-Hill method, it is helpful to have a table of cutoff values handy. Table 1-4 gives cutoff values (rounded to three decimal places) for quotas between 1 and 15.

TABLE 1-4			
Lower quota L	Upper quota U	Huntington-Hill cutoff G	Webster cutoff A
1	2	1.414	1.5
2	3	2.449	2.5
3	4	3.464	3.5
4	5	4.472	4.5
5	6	5.477	5.5
6	7	6.481	6.5
7	8	7.483	7.5
8	9	8.485	8.5
9	10	9.487	9.5
10	11	10.488	10.5
11	12	11.489	11.5
12	13	12.490	12.5
13	14	13.491	13.5
14	15	14.491	14.5

Table 1-4 illustrates how and where the two sets of rounding rules differ. For example, a quota of 3.48 would be *rounded up under the Huntington-Hill rule* (it's above the cutoff point of 3.464), but rounded down under Webster's rule. On the other hand, a quota of 14.48 would be rounded down under both rules. The impression we get, looking carefully at Table 1-4, is that there is only a very small "window of opportunity" for the two rounding rules to act differently, and that as the quotas get larger, this small window gets even smaller. This is indeed what happens.

We will now illustrate the difference between Webster's method and the Huntington-Hill method with a pair of examples.

See Exercise 18.

❯ EXAMPLE 1.5 Huntington-Hill Can Be Easier Than Webster

A small country consists of five states A, B, C, D, and E, with populations of 34,800 (A); 104,800 (B); 64,800 (C); 140,800 (D); and 54,800 (E). There are $M = 40$ seats in the legislature. We will apportion the seats first using the Huntington-Hill method, and then, for comparison purposes, using Webster's method.

Table 1-5 shows the first set of calculations, using the standard divisor and standard quotas. The third row of the table gives the standard quotas, based on the standard divisor $SD = 400,000/40 = 10,000$. The fourth row shows the results of rounding the standard quotas using the Huntington-Hill rounding rules. Notice that under the Huntington-Hill rounding rules the quotas 3.48 and 5.48 get *rounded up* (they are both above their cutoff points).

TABLE 1-5

State	A	B	C	D	E	Total
Population	34,800	104,800	64,800	140,800	54,800	400,000
Standard quota	3.48	10.48	6.48	14.08	5.48	40
Huntington-Hill	**4**	**10**	**6**	**14**	**6**	**40**
Webster	3	10	6	14	5	38

The nice surprise is that using the Huntington-Hill rounding rules we get an apportionment—the standard divisor is just the right divisor to apportion under the Huntington-Hill method! On the other hand, we can see from the last row that the standard quotas don't work for Webster's method, as 3.48 and 5.48 get rounded down.

Table 1-6 shows a set of modified quotas that works for Webster's method. They were obtained using the divisor $D = 9965$. (It takes a few educated guesses to hit on this divisor. Other divisors between 9964 and 9969 also work.) The last row shows the apportionment under Webster's method.

TABLE 1-6

State	A	B	C	D	E	Total
Population	34,800	104,800	64,800	140,800	54,800	400,000
Quota ($D = 9965$)	3.492	10.517	6.503	14.13	5.499	
Webster	3	**11**	**7**	**14**	5	40

The next example is, in a sense, the mirror image of Example 1.5. The background story is the same, with only a few minor changes in the population figures.

▶ EXAMPLE 1.6 Huntington-Hill Can Be Harder Than Webster

A small country consists of five states A, B, C, D, and E, with populations of 34,800 (A); 105,100 (B); 65,100 (C); 140,200 (D), and 54,800 (E). There are $M = 40$ seats in the legislature. We will apportion the seats using both the Huntington-Hill method and Webster's method.

Table 1-7 shows the first set of calculations based on the standard divisor $SD = 400,000/40 = 10,000$. The third row of the table gives the standard quotas. The fourth row shows the results of rounding the standard quotas using the Huntington-Hill rounding rules, and the last row shows the results of rounding the quotas using Webster's rounding rules. We can see that in this case the standard divisor $SD = 10,000$ is a suitable divisor to apportion under Webster's method but is not a suitable divisor for the Huntington-Hill method!

TABLE 1-7

State	A	B	C	D	E	Total
Population	34,800	105,100	65,100	140,200	54,800	400,000
Standard quota	3.48	10.51	6.51	14.02	5.48	40
Huntington-Hill	4	11	7	14	6	42
Webster	3	11	7	14	5	40

A suitable divisor for the Huntington-Hill method will have to be a number bigger than 10,000 (we need to bring the standard quotas down a bit). A little nip here, a little tuck there, and we find $D = 10,030$, a divisor that works! Table 1-8 shows the quotas when we use $D = 10,030$. When we carefully round these quotas using the Huntington-Hill rounding rules (good idea to have Table 1-4 handy!) we get the Huntington-Hill apportionment shown in the last row of Table 1-8.

TABLE 1-8

State	A	B	C	D	E	Total
Population	34,800	105,100	65,100	140,200	54,800	400,000
Quota ($D = 10,030$)	3.47	10.479	6.491	13.978	5.464	
Huntington-Hill	4	10	7	14	5	40

We conclude this discussion with a note of caution. Looking at the results of Examples 1.5 and 1.6, we might get the impression that the Huntington-Hill method and Webster's method are likely to produce different apportionments, but this is far from being the case. The numbers in Examples 1.5 and 1.6 were carefully rigged to show that the two methods *can* give different results—we should hardly conclude that this happens all the time. In fact, in a real-life example the odds are pretty high that the same divisor D will work for both Webster and Huntington-Hill apportionments.

Since the implementation of the Huntington-Hill method in 1941, every apportionment of the House of Representatives would have turned out exactly the same had Webster's method been used.

Conclusion

The Huntington-Hill method is of interest primarily for a practical reason—it is by law, the way the United States has chosen (after trying several other methods) to apportion its House of Representatives. The method was chosen as the "best" of all possible apportionment methods based on mathematical considerations that go beyond the scope of our discussion, but as a practical matter, the method is almost identical to Webster's method (and most of the time produces exactly the same results). As educated citizens, it behooves us to have some basic understanding of how the system works, including how is it that California has 53 seats in the House of Representatives, but Arkansas has only 4. Unfortunately, the mathematics behind the Huntington-Hill method are a bit above and beyond the level that most people are comfortable with, but to us, veterans of the Chapter 4 expeditions, this mini-excursion was just a natural extension of that trip. You should derive some sense of satisfaction and pat yourself on the back for having completed this journey.

But first, try some of the exercises.

Exercises

A. The Geometric Mean

1. Without using a calculator, find the geometric mean of each pair of numbers.

   (a) 10 and 1000

   (b) 20 and 2000

   (c) 1/20 and 1/2000

2. Without using a calculator, find the geometric mean of each pair of numbers.

   (a) 2 and 8

   (b) 20 and 80

   (c) 1/20 and 1/80

3. Without using a calculator, find the geometric mean of $2 \cdot 3^4 \cdot 7^3 \cdot 11$ and $2^7 \cdot 7^5 \cdot 11$.

   (***Hint:*** *These are very big numbers, but you don't need a calculator—work with their prime factorizations.*)

4. Without using a calculator, find the geometric mean of $11^5 \cdot 13^8 \cdot 17^3 \cdot 19$ and $11^7 \cdot 17^5 \cdot 19$.

   (***Hint:*** *These are very big numbers, but you don't need a calculator—work with their prime factorizations.*)

5. Home prices in Royaltown increased by 11.7% in 2005 and 25.9% in 2006. Find the average annual increase in home prices over the two-year period (rounded to the nearest tenth of a percent).

6. Over a two-year period a company's shares went up by 7.25% the first year and 11.45% the second year. Find the average annual increase in the value of the shares over the two-year period (rounded to the nearest tenth of a percent).

7. Using a calculator, complete the following table. Round the answer to three decimal places.

Consecutive integers	Geometric mean (*G*)	Arithmetic mean (*A*)	Difference (*A* − *G*)
15, 16			
16, 17			
17, 18			
18, 19			
19, 20			
29, 30			
39, 40			
49, 50			

8. Let *a* and *b* be positive numbers, and let *G* be their geometric mean. Express the geometric mean of each of the following pairs of numbers in terms of *G*.

   (a) $3.75a$ and $3.75b$

   (b) $10a$ and $1000b$

   (c) $1/a$ and $1/b$

   (d) a^2 and b^2

9. Let *a* and *b* be two positive numbers, let *G* be their geometric mean, and let *k* be a positive constant. Show that

   (a) The geometric mean of $k \cdot a$ and $k \cdot b$ is $k \cdot G$.

   (b) The geometric mean of k/a and k/b is k/G.

10. The arithmetic-geometric mean inequality. Let a and b be positive numbers, let A be their arithmetic mean, and let G be their geometric mean. Show that

(a) If $b = a$, then $A = G$.

(b) If $b > a$, then $A > G$.

(**Hint:** *Show that $a^2 + b^2 \geq 2 \cdot a \cdot b$, and use this to show that $\left(\dfrac{a + b}{2} \right)^2 \geq a \cdot b$.*)

11. Pythagoras and the arithmetic-geometric mean inequality. Let a and b be positive numbers, and $b > a$.

(a) In the right triangle ABC shown in the figure, find the length of BC.

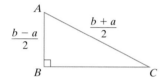

(b) Explain why the result of (a) proves the arithmetic-geometric mean inequality.

B. The Huntington-Hill Method

12. Round each quota using the Huntington-Hill rounding rules.

(a) $q = 4.46$

(b) $q = 4.48$

(c) $q = 50.498$

(d) $q = \sqrt{12.01}$

13. A small country consists of five states: A, B, C, D, and E. The standard quotas of each state are given in the following table. Find the apportionment under the Huntington-Hill method.

State	A	B	C	D	E
Standard quota	25.26	18.32	2.58	37.16	40.68

14. (a) Use the Huntington-Hill method to apportion Parador's Congress (Chapter 4, Example 4.3).

(b) Compare your answer in (a) with the apportionment produced by Webster's method. What's your conclusion?

15. A country consists of six states, with the state's populations given in the following table. The number of seats to be apportioned is $M = 200$.

State	A	B	C	D	E	F
Population	344,970	408,700	219,200	587,210	154,920	285,000

(a) Find the apportionment under Webster's method.

(b) Find the apportionment under the Huntington-Hill method.

(c) Compare the apportionments found in (a) and (b).

16. A country consists of six states, with the state's populations given in the following table. The number of seats to be apportioned is $M = 200$.

State	A	B	C	D	E	F
Population	344,970	204,950	515,100	84,860	154,960	695,160

(a) Find the apportionment under Webster's method.

(b) Find the apportionment under the Huntington-Hill method.

(c) Compare the apportionments found in (a) and (b).

17. The small island nation of Margarita consists of four islands: Aleta, Bonita, Corona, and Doritos. The state's population of each island is given in the following table. The number of seats to be apportioned is $M = 100$.

State	A	B	C	D
Population	86,915	4,325	5,400	3,360

(a) Find the apportionment under the Huntington-Hill method.

(b) Describe any possible violations that occurred under the apportionment in (a).

C. Miscellaneous

18. Show that the difference between the arithmetic and geometric means of consecutive positive integers gets smaller as the integers get bigger.

(**Hint:** *You should show that for any positive integer N, the difference between the arithmetic and geometric means of N and N + 1 is bigger than the difference between the arithmetic and geometric means of N + 1 and N + 2.*)

19. Let a, b, and c be positive numbers such that $a < b < c$. Let A denote the arithmetic mean of a and b, and A' denote the arithmetic mean of a and c. Likewise, let G denote the geometric mean of a and b, and G' denote the geometric mean of a and c. Show that $A' - G' > A - G$.

(**Hint:** *Start by showing that $\sqrt{c} - \sqrt{a} > \sqrt{c} - \sqrt{b}$ and then square both sides of the inequality.*)

20. The purpose of this exercise is to show that under rare circumstances, there is no suitable divisor for the Huntington-Hill method, and thus the method will not work. (Imagine the headache this scenario would create in the U.S. Congress!) A small country consists of three states, with populations given in the following table. The number of seats to be apportioned is $M = 12$.

State	A	B	C
Population	7290	1495	1215

(a) Show that using the Huntington-Hill rounding rules with the divisor $D = \dfrac{1215}{\sqrt{2}}$ does not work. Explain why this divisor is too small!

(**Hint:** *Do not convert the numbers to decimal form—do all your computations using radicals.*)

(b) Show that using the Huntington-Hill rounding rules any divisor $D > \dfrac{1215}{\sqrt{2}}$ does not work. Explain why any divisor of this size is too big!

(c) Explain why (a) and (b) imply that the Huntington-Hill method will not work in this apportionment problem.

21. The quadratic mean. The *quadratic mean* of two numbers a and b is defined as $Q = \sqrt{\dfrac{a^2 + b^2}{2}}$.

(a) Using a calculator, complete the following table. Round the answer to three decimal places.

a	b	Q	a	b	Q
1	2		9	10	
2	3		19	20	
3	4		29	30	
4	5		99	100	
5	6		10	20	
6	7		10	90	
7	8		10	190	
8	9		10	990	

(b) Let a and b be positive numbers such that $a < b$. Show that $a < Q < b$.

(c) Let a and b be positive numbers such that $a < b$. Show that the quadratic mean of a and b is bigger than the arithmetic mean of a and b ($Q > A$).

22. The harmonic mean. The *harmonic mean* of two numbers a and b is defined as $H = \dfrac{2a \cdot b}{a + b}$.

(a) Using a calculator, complete the following table. Round the answer to three decimal places.

a	b	H	a	b	H
1	2		9	10	
2	3		19	20	
3	4		29	30	
4	5		99	100	
5	6		10	20	
6	7		10	90	
7	8		10	190	
8	9		10	990	

(b) Let a and b be positive numbers such that $a < b$. Show that $a < H < b$.

(c) Let a and b be positive numbers such that $a < b$. Show that the harmonic mean of a and b is smaller than the geometric mean of a and b ($H < G$).

Projects and Papers

A. Dean's Method

Dean's method is a method almost identical to the Huntington-Hill method (and thus to Webster's method) that uses the *harmonic means* of the lower and upper quotas (see Exercise 22) as the cutoff points for rounding. In this project you are to discuss Dean's method by comparing it to the Huntington-Hill method (in a manner similar to how the Huntington-Hill method was discussed in the Mini Excursion by comparison to Webster's method). Give examples that illustrate the difference between Dean's method and the Huntington-Hill method. You should also include a brief history of Dean's method.

B. *Montana v. U.S. Department of Commerce* (U.S. District Court, 1991); *U.S. Department of Commerce v. Montana* (U.S. Supreme Court, 1992)

Write a paper discussing these two important legal cases concerning the apportionment of the U.S. House of Representatives. In this paper you should (i) present the background preceding Montana's challenge to the constitutionality of the Huntington-Hill method, (ii) summarize the arguments presented by Montana and the government in both cases, (iii) summarize the arguments given by the District Court in ruling for Montana, and (iv) summarize the arguments given by the Supreme Court in unanimously overturning the District Court ruling.

Appendix 2000–2010 Apportionments of the House of Representatives

TABLE 1-9 **Current Huntington-Hill Apportionments**

State	Population	Seats	State	Population	Seats
Alabama	4,461,130	7	Montana	905,316	1
Alaska	628,933	1	Nebraska	1,715,369	3
Arizona	5,140,683	8	Nevada	2,002,032	3
Arkansas	2,679,733	4	New Hampshire	1,238,415	2
California	33,930,798	53	New Jersey	8,424,354	13
Colorado	4,311,882	7	New Mexico	1,823,821	3
Connecticut	3,409,535	5	New York	19,004,973	29
Delaware	785,068	1	North Carolina	8,067,673	13
Florida	16,028,890	25	North Dakota	643,756	1
Georgia	8,206,975	13	Ohio	11,374,540	18
Hawaii	1,216,642	2	Oklahoma	3,458,819	5
Idaho	1,297,274	2	Oregon	3,428,543	5
Illinois	12,439,042	19	Pennsylvania	12,300,670	19
Indiana	6,090,782	9	Rhode Island	1,049,662	2
Iowa	2,931,923	5	South Carolina	4,025,061	6
Kansas	2,693,824	4	South Dakota	756,874	1
Kentucky	4,049,431	6	Tennessee	5,700,037	9
Louisiana	4,480,271	7	Texas	20,903,994	32
Maine	1,277,731	2	Utah	2,236,714	3
Maryland	5,307,886	8	Vermont	609,890	1
Massachusetts	6,355,568	10	Virginia	7,100,702	11
Michigan	9,955,829	15	Washington	5,908,684	9
Minnesota	4,925,670	8	West Virginia	1,813,077	3
Mississippi	2,852,927	4	Wisconsin	5,371,210	8
Missouri	5,606,260	9	Wyoming	495,304	1
			Total	**281,424,177**	**435**

References and Further Readings

1. Balinski, Michel L., and H. Peyton Young, *Fair Representation; Meeting the Ideal of One Man, One Vote.* New Haven, CT: Yale University Press, 1982.

2. Census Bureau Web site, *http://www.census.gov.*

3. Huntington, E. V., "The Apportionment of Representatives in Congress." *Transactions of the American Mathematical Society,* 30 (1928), 85–110.

4. Huntington, E. V., "The Mathematical Theory of the Apportionment of Representatives," *Proceedings of the National Academy of Sciences, U.S.A.,* 7 (1921), 123–127.

5. Saari, D. G., "Apportionment Methods and the House of Representatives," *American Mathematical Monthly,* 85 (1978), 792–802.

6. Schmeckebier, L. F., *Congressional Apportionment.* Washington, DC: The Brookings Institution, 1941.

A Touch of Color

Graph Coloring

FIGURE 2-1

The map in the figure above is clearly a map of the 48 lower states of the United States, but it lacks a little charm. Maps are supposed to be visually pleasing, and the typical way to make them so is to add a touch of color. The standard approach to coloring a map is to use a single color for a state (we will use "state" as a metaphor for the geographical units in the map, which in reality could be countries, provinces, etc.), and never use the same color for two states that share a common border. On the other hand, two states whose common border is just one point (for example Arizona and Colorado) can be colored, if we so choose, with the same color.

The colorful map shown in Fig. 2-2(a) is a bit too colorful: 48 colors were used, one for each state. First, it's a little jarring to see that many colors—a classic case of too much of a good thing. Second, let's imagine that there is a production cost incurred each time we add another color to the map, so that a map that uses say 48 colors is more expensive than a map that uses 47 colors, which in turn is more expensive than a map that uses 46 colors, and so on. (In the old days this was probably true, but with modern laser and inkjet printers the argument doesn't hold up. On the other hand, the cost argument gives us a convenient way to think about the issues in map coloring, so we will pretend it's still true.) A more typical, "less expensive" colored map is shown in Fig. 2-2(b). Here a total of five different colors are used. Is it possible to cut down the number of colors used to four? How about three? You are encouraged to try answering these questions before you read on. (For a convenient way to experiment, go to *http://www.sailor.lib.md.us/MD_topics/kid/col_applet/color.html*. If you choose "US Map" from the drop down menu that shows "Flag," and then click on "Load Image" you will get a blank map of the lower 48 states and a palette of colors for coloring the map.)

See Exercise 16.

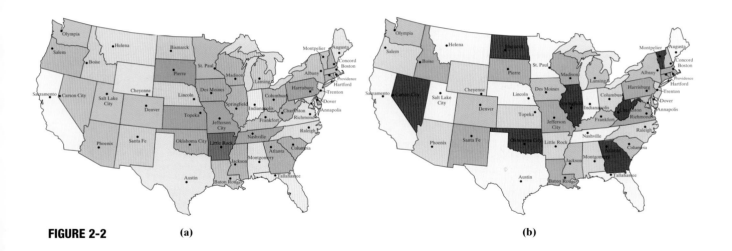

FIGURE 2-2   **(a)**   **(b)**

Much has been written about the mathematics of map coloring and its connection with graph theory. In this mini-excursion we will explore some of the ideas behind the fascinating topic of graph coloring and some of its more colorful applications. Some familiarity with the basic terminology and concepts in Chapters 5 and 6 is strongly recommended.

Graph Coloring and Chromatic Numbers

▶ EXAMPLE 2.1 Can't We All Get Along?

Sometimes people just don't get along, and you are caught in the middle. Imagine that you are a wedding planner organizing the rehearsal dinner before a big wedding. There are a total of 16 people attending the rehearsal dinner: $A, B, C, \ldots H$ are relatives of the bride and groom; $I, J, K, \ldots P$ are members of the wedding party. If things weren't stressful enough, you are told that some of these people have serious issues:

- *A* doesn't get along with *F*, *G*, or *H*,
- *B* doesn't get along with *C*, *D*, or *H*,
- *C* doesn't get along with *B*, *D*, *E*, *G*, or *H*,
- *D* doesn't get along with *B*, *C*, or *E*,
- *E* doesn't get along with *C*, *D*, *F*, or *G*,
- *F* doesn't get along with *A*, *E*, or *G*,
- *G* doesn't get along with *A*, *C*, *E*, or *F*,
- *H* doesn't get along with *A*, *B*, or *C*.

To make the rehearsal dinner go smoothly you are instructed to find a way to seat these people so that people that don't get along must be seated at different tables. (*I* through *P* get along with everyone, so they are not a concern.) How are you going to set up the seating arrangements with so many incompatibility issues to worry about? What is the minimum number of tables you will need? You can answer both of these questions using a little graph theory.

We will start by creating the "incompatibility" graph shown in Fig. 2-3(a). In this graph the vertices represent the individuals, and two vertices are connected by an edge if the corresponding individuals don't get along (and therefore, should not be seated at the same table). To make seating assignments we assign colors to the vertices of the graph, with each color representing a table. Since we don't want two incompatible individuals seated at the same table, *we don't want to color two vertices that are connected by an edge with the same color*. We will refer to any coloring that satisfies this rule as a *legal coloring* of the graph.

Figure 2-3(b) shows a legal coloring of the vertices of the graph in Fig. 2-3(a) that uses four different colors. This coloring tells us how to seat this obnoxious group using four tables. Can we do better (i.e. use less than four tables)? Yes. Figure 2-3(c) shows a legal coloring of the vertices of the graph that uses just three colors.

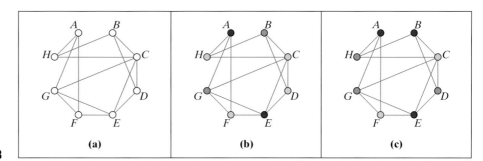

FIGURE 2-3

Could we possibly come up with a legal coloring of the vertices of the graph that uses just two colors? No. The reason is clear if we look at *A*, *F*, and *G* (or any other set of three vertices that form a *triangle*). Since each is adjacent to the other two, *A*, *F*, and G will have to be colored with different colors.

The conclusion to our analysis is: (i) the minimum number of tables needed to sit the wedding party is three, and (ii) the seating assignment should put *A*, *B*, and *E* in one table (red), *C* and *F* in a second table (blue), and *D*, *G*, and *H* in the third table (green). The remaining members of the wedding party can be arbitrarily assigned to fill up the remaining seats at the three tables.

This is, in fact, the unique solution to the problem—see Exercise 12.

Example 2.1 illustrates the two key ideas of this mini-excursion: (i) the idea of coloring the vertices of a graph so that *adjacent vertices don't get the same color*, and (ii) the concept of trying to do this using as *few colors as possible*. Here are the formal definitions of these concepts.

k-coloring

A ***k*-coloring** of a graph G is a coloring of the vertices of G using k colors and satisfying the requirement that adjacent vertices are colored with different colors.

Chromatic Number

The **chromatic number** of a graph G is the *smallest* number k for which a k-coloring of the vertices of G is possible. We will use the notation $\chi(G)$ to denote the chromatic number of G.

Note that the actual choice of colors used in coloring the graph is irrelevant. In fact, the colors are just a convenient way to classify the vertices, and we could just as well use numbers or letters to do the same. The use of colors is based primarily on the fact that humans are better able to perceive patterns of colors than patterns of symbols.

Using the above notation, for the graph G in Fig. 2-3(a) we have $\chi(G) = 3$. This follows from the observation that a 3-coloring of G is possible, as shown in Fig. 2-3(c), but a 2-coloring of G is not.

► EXAMPLE 2.2 Coloring Complete Graphs

The graph in Fig. 2-4(a) is K_5, the complete graph on 5 vertices. In this graph every vertex is adjacent to every other vertex, so no two vertices can have the same color. The only possible way to color K_5 is to use a different color for each vertex, as in Fig. 2-4(b). Thus, we can conclude that $\chi(K_5) = 5$.

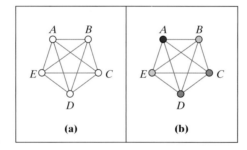

FIGURE 2-4

The argument used in Example 2.2 can be generalized to show that $\chi(K_n) = n$. This illustrates the fact that it is possible to create graphs with arbitrarily large chromatic numbers. You need a graph G with $\chi(G) = 100$? No problem—choose G to be K_{100}.

► EXAMPLE 2.3 Coloring Circuits

A graph consisting of a single circuit with n vertices is denoted by C_n. The graph shown in Fig. 2-5(a) is C_6. Figure 2-5(b) shows a 2-coloring of C_6. Since a 1-coloring is clearly out of the question, we can conclude that $\chi(C_6) = 2$.

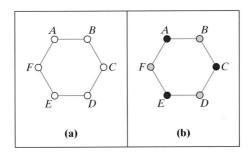

FIGURE 2-5

(a)   (b)

The graph shown in Fig. 2-6(a) is C_5. Figure 2-6(b) shows a 3-coloring of C_5. A little reflection should be enough to convince ourselves that we will not be able to color C_5 with just two colors, as we did with C_6. The problem is that we can alternate two colors around the circuit until we get to the last vertex, but since the number of vertices is odd, the last vertex will be adjacent to two vertices of different colors, and a third color will be needed. It follows that $\chi(C_5) = 3$.

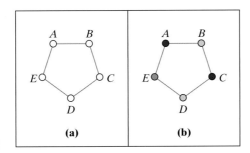

FIGURE 2-6

(a)   (b)

See Exercise 9.

We can generalize the preceding observations to any $C_n (n \geq 3)$: If n is even, $\chi(C_n) = 2$; if n is odd, $\chi(C_n) = 3$. Thus, we now know that it is possible to have graphs with a large number of vertices and small chromatic number (2 or 3). Graphs with a large number of vertices and chromatic number 1 are also possible, but are extremely uninteresting—they have no edges.

See Exercise 8.

The Greedy Algorithm for Graph Coloring

Like some of the other graph problems discussed in Chapters 5 through 8 (Euler circuit problems, traveling salesman problems, shortest network problems, scheduling problems) graph coloring can be thought of as an *optimization* problem: How can we color a graph using the fewest possible number of colors? We will call such a coloring an *optimal coloring* of the graph, and the general problem of finding optimal colorings is known as the *coloring problem*.

> **Optimal Coloring**
>
> An **optimal coloring** of a graph G is a coloring of the vertices of G using the fewest possible number of colors. To put it in slightly more formal terminology, an optimal coloring of G is a $\chi(G)$-coloring of G. [Too many G's in the last sentence, but all it says is that when we color a graph using $\chi(G)$ colors by definition we have an optimal coloring of the graph.]

The truly interesting question in graph coloring is the following: Given an arbitrary graph G, how do we find an optimal coloring of G? If you are a veteran of Chapters 5 through 8, you may not be entirely surprised to find out that *no efficient general algorithm is known for finding optimal colorings of graphs*. For small graphs we can use trial and error, and for some families of graphs optimal coloring is reasonably easy (for example complete graphs, circuits, etc.) but those are just special cases. In general, the best we can do is to use *approximate algorithms* that hopefully get us close to an optimal coloring.

In this section we will discuss a classic approximate algorithm for graph coloring known as the *greedy algorithm* (for reasons that will become clear soon). To illustrate the ideas behind the greedy algorithm we will start with a couple of simple examples.

EXAMPLE 2.4 Greed Is Good (Sometimes)

Let's try to color the graph in Fig. 2-7(a). The strategy we will use is simple. We will start with vertex A and color it with some color, say blue. We will think of blue as the *first* color in some arbitrary priority list of colors (blue goes first, red second, green third, yellow fourth, and so on.) We now move to the next vertex, B and try to color it with the first color in our list (blue), but we can't do it because B is adjacent to A, which is already blue. OK, fine—we'll just go to the next available color on the list (red) and use it for B. If we go next to vertex C how should we color it? Other than red (C is adjacent to B) we can use any color we want, but why introduce a third color when we can use blue? We are, after all, trying to cut down on the number of colors used. So we color C with blue. Using the same philosophy (try the colors in the designated order) we color D with red, and so on. Everything lines up just right and we can alternate blue and red and get the 2-coloring shown in Fig. 2-7(b). Since we know that the graph cannot be colored with one color, Fig. 2-7(b) gives an optimal 2-coloring of the graph.

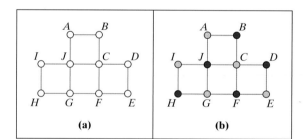

FIGURE 2-7

The strategy we used in Example 2.4 has two key components: (i) we color the vertices of the graph following some designated order (in Example 2.4 we went through the vertices in alphabetical sequence, but it can be done in any order we choose) and (ii) we have a priority list for assigning the colors, and we limit the colors used by always starting at the top of the list and using previously used colors as much as possible. For obvious reasons, we call this a *greedy strategy*.

In Example 2.4 the greedy strategy gave us an optimal coloring, but there was a bit of good karma involved. By pure luck, the order in which we colored the vertices just worked out perfectly. This is not always the case, and our next example shows how bad luck in the order of the vertices can give us a bad coloring of the graph.

> ### EXAMPLE 2.5 Greed Can Sometimes Be Bad

The graph in Fig. 2-8(a) is the same graph as the one in Fig. 2-7(a) except for the way the vertices are labeled. If we try the same approach we used in Example 2.4 (color the vertices in alphabetical order and use the same priority order for colors—blue first, red second, green third, yellow fourth) we will get a very different result. We start by coloring vertex A with blue. So far so good. Now B is not adjacent to A, so we will also color it with blue. Ditto for C and D [Fig. 2-8(b)]. So far we have used only one color, so we are doing well. When we get to vertex E we have to go down to our second color because E is adjacent to D, which has already been colored blue. Thus, E is colored red. Likewise, F is adjacent to a blue vertex, so we color it red. We are now looking at Fig. 2-8(c). Since G is now adjacent to a blue and a red vertex, we have to pull out a third color (green) for G [Fig. 2-8(d)]. Now you can clearly see our bad luck. The next vertex (H) is adjacent to a blue, a red, and a green vertex so we need a fourth color (yellow) to color it [Fig. 2-8(e)]. The last two vertices (I and J) are adjacent to blue vertices only, so we will color them red. This gives us the final coloring of the graph [Fig. 2-8(f)].

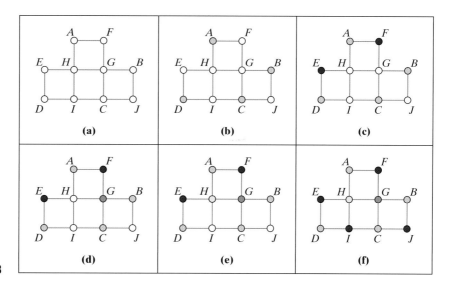

FIGURE 2-8

See Exercise 21.

Examples 2.4 and 2.5 illustrate an important point: The order in which the vertices of the graph are colored can make a big difference in the kind of coloring we get. There is always some way to order the vertices so that we get an optimal coloring, but there is no easy way to figure out what that order is. Moreover, there are $n!$ different ways to order the n vertices, so going through the different orderings to find the best one is out of the question. (Take a quick look at Table 6-4 in Chapter 6 if you need to jog your memory on how quickly factorials grow.)

We will now formalize the ideas introduced in Examples 2.4 and 2.5 into what is known as the *greedy algorithm* for graph coloring. To implement the algorithm we assume that the vertices are ordered in some arbitrary order $v_1, v_2, \ldots, v_n$, and that $c_1, c_2, c_3, c_4, \ldots$ represents a priority order for the colors.

Greedy Algorithm for Graph Coloring

- **Step 1.** Assign the first color (c_1) to the first vertex (v_1).
- **Step 2.** Vertex v_2 is assigned color c_1 if it is not adjacent to v_1; otherwise it gets assigned color c_2.
- **Steps 3, 4, . . . , *n*.** Vertex v_i is assigned the first possible color in the priority list of colors (i.e. the first color that has not been assigned to one of the already colored neighbors of v_i).

In spite of its limitations, the greedy algorithm is a reasonable approach for graph coloring, especially when we use some ingenuity in the way we order the vertices. Since there are more restrictions in coloring vertices of higher degree than there are in coloring vertices of lower degree, one obvious refinement of the greedy algorithm is to order the vertices in *decreasing* order of degrees: color vertices of highest degrees first (if there are ties choose among them at random), color vertices of next highest degree next, and so on. If we were to apply this approach to the graph in Fig. 2-8 one possible ordering of the vertices would be: G, $H, C, I, A, B, D, E, F, J$. If we were to color the vertices in this order using the greedy algorithm we would get an optimal coloring of the graph in Fig. 2-8.

See Exercise 20.

The greedy algorithm can also be used to produce upper bounds on the chromatic number of the graph. The best known of these upper bounds is given by the following fact, known as Brook's Theorem.

Brook's Theorem

Let $d_1 \geq d_2 \geq \cdots \geq d_n$ be the degrees of the vertices of the graph G listed in decreasing order. Then $\chi(G) \leq d_1 + 1$.

At most d_1 colors

FIGURE 2-9

Brook's Theorem essentially says that *the chromatic number of a graph cannot be more than the largest degree in the graph plus one*. To see why this is true, think about the worst case scenario we can run into when we are coloring a vertex using the greedy algorithm: the vertex is adjacent to many other vertices all of which have been colored with different colors. Since the largest number of adjacent vertices a vertex can have is d_1, the worst thing that could happen is that the first d_1 colors have been used and we would need one more to color our vertex (Fig. 2-9).

It turns out that for *connected graphs*, the only two cases where we actually have to use the maximum $d_1 + 1$ colors to color the graph are when the graph is the complete graph K_n (all vertices have degree $n - 1$, the chromatic number is n), or the circuit C_n where n is odd (all vertices have degree 2, the chromatic number is 3). If we rule these two cases out, we no longer need the $+1$ in Brook's Theorem. We will call this the *strong version* of Brook's Theorem.

Brook's Theorem (Strong Version)

Let $d_1 \geq d_2 \geq \cdots \geq d_n$ be the degrees of the vertices of a *connected* graph G listed in decreasing order. If G is not C_n (n odd) or K_n, then $\chi(G) \leq d_1$.

Graph Coloring and Sudoku

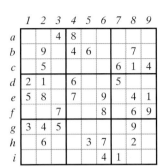

FIGURE 2-10

The game of Sudoku has become, in the last few years, the rage among puzzle and game enthusiasts looking for a more intellectual (and cheaper) challenge than the one provided by an X-Box. Sudoku is addictive, and even ordinary people that are not drawn to video games are hooked on it. These days practically every major newspaper carries a daily Sudoku puzzle.

If you haven't played Sudoku yet, the rules are quite simple: You start with a 9×9 grid of 81 squares called *cells*. The grid is also subdivided into nine 3×3 subgrids called *boxes*. Some of the cells are already filled with the numbers 1 through 9. These are called the *givens*. The challenge of the game is to complete the grid by filling the remaining cells with the numbers 1 through 9. The requirements are: (i) every row and every column of the grid must have the numbers 1 through 9 appear once; (ii) each of the nine boxes must have the numbers 1 through 9 appear once.

A typical Sudoku puzzle may have somewhere between 25 and 40 givens, depending on the level of difficulty. Figure 2-10 is an example of a moderately easy Sudoku puzzle. (*Source*: *http://www.geometer.org/mathcircles*). The labels 1 through 9 on the columns and *a* through *i* on the rows are not part of the puzzle, but they provide a convenient way to refer to the cells. Just for fun, you may want to try this one out before you read on. (*Hint*: Try to figure out what number should go in cell $f2$. Once you have that one figured, go to cell $d3$. That's enough help for now. The solution is shown after the References and Further Readings.)

The Sudoku Graph

To see the connection between Sudoku and graph coloring, we will first describe the **Sudoku graph**, which for convenience we will refer to as S. The graph S has 81 vertices, with each vertex representing a cell. When two cells *cannot* have the same number (either because they are in the same row, in the same column, or in the same box) we put an edge connecting the corresponding vertices of the Sudoku graph S. For example, since cells $a3$ and $a7$ are in the same row, there is an edge joining their corresponding vertices; there is also an edge connecting $a1$ and $b3$ (they are in the same box), and so on. When everything is said and done, each vertex of the Sudoku graph has degree 20, and the graph has a total of 810 edges. S is too large to draw, but we can get a sense of the structure of S by looking at a partial drawing such as the one in Fig. 2-11. The drawing shows all 81 vertices of S, but only two ($a1$ and $e5$) have their full set of incident edges showing.

See Exercise 22.

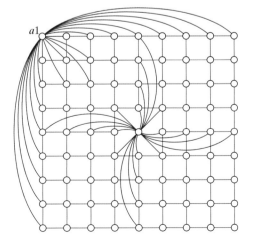

FIGURE 2-11 A partial drawing of the Sudoku graph

The second step in converting a Sudoku puzzle into a graph coloring problem is to assign colors to the numbers 1 through 9. This assignment is arbitrary, and is not a priority ordering of the colors as in the greedy algorithm —it's just a simple correspondence between numbers and colors. Figure 2-12 shows one such assignment.

FIGURE 2-12

Cell number:	1	2	3	4	5	6	7	8	9
Vertex color:	●	○	○	○	○	○	○	●	○

Once we have the Sudoku graph and an assignment of colors to the numbers 1 through 9, any Sudoku puzzle can be described by a Sudoku graph where some of the vertices are already colored (the ones corresponding to the givens). For example, the Sudoku puzzle shown in Fig. 2-10 is equivalent to the partial coloring shown in Fig. 2-13. To solve the Sudoku puzzle all we have to do is color the rest of the vertices using the nine colors in Fig. 2-12.

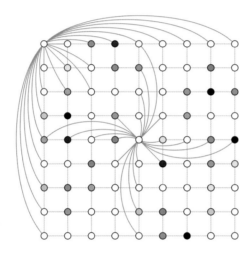

FIGURE 2-13 A Sudoku puzzle as a graph coloring problem

Map Coloring

Map drawing and coloring is an ancient art, but the connection between map coloring and mathematics originated in 1852 when a University of London student by the name of Francis Guthrie mentioned to his mathematics professor (the well known mathematician Augustus De Morgan) that he had been coloring many maps of English counties (don't ask why) and noticed that every map he had tried, no matter how complicated, could be colored with four colors (where districts with a common border had to be colored with different colors). He inquired if this was a known mathematical fact, and De Morgan (who didn't know the answer) checked with some of the most famous mathematicians of the time, including his friend William Rowan Hamilton.

A short biographical profile of Hamilton is given in Chapter 6.

Guthrie's notion that any map could be colored with just four colors, sounded so simple that everyone assumed it could be easily proved mathematically. After 100 years and many failed attempts at a proof, the *Four-Color Conjecture*, as the problem was famously known, was finally solved in 1976 by Kenneth Appel and Wolfgang Haken of the University of Illinois. Yes, indeed, Guthrie was right: Every map can be colored with four colors or less. The solution to this simple question took up 500 pages and about 1000 hours of computer time.

Any map coloring problem can be converted into a graph coloring problem by first finding the **dual graph** of the map. In the dual graph, each vertex represents a "state" (remember that this is a metaphor for the political units used in the map), and two vertices are connected by an edge if the corresponding states have a common border. If the common border happens to be just a point (as is the case with Arizona and Colorado), then we do not connect the vertices. The problem of coloring the map using the fewest number of colors now becomes simply the problem of *finding an optimal coloring of the vertices of the dual graph.* We can now take all the tools and techniques we learned for graph coloring and use them for map coloring.

Our final example is a very small example whose only purpose is to illustrate the idea behind coloring a map by means of its dual graph.

For a slightly more meaningful map coloring problem, see Exercise 15.

> **EXAMPLE 2.6** Map Coloring with Dual Graphs

Figure 2-14(a) shows a small map with six states, and Fig. 12-14(b) shows its dual graph. Figure 2-14(c) shows an optimal 3-coloring of the dual graph. We know that the 3-coloring is optimal because the graph has triangles, and thus cannot be colored with 2 colors. The corresponding optimal coloring of the map is shown in Fig. 2-14(d).

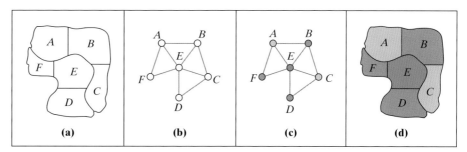

FIGURE 2-14

(a)   (b)   (c)   (d)

Exercises

A. Graph Colorings and Chromatic Numbers

For Exercises 1 through 6, you can experiment with coloring the graphs using a Java applet available at: http://www.cut-the-knot.org/Curriculum/Combinatorics/ColorGraph.shtml. The applet was designed specifically to match these exercises.

1. For the graph G shown in the figure,

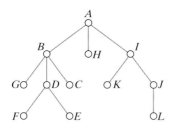

(a) find a 3-coloring of G.

(b) find a 2-coloring of G.

(c) find $\chi(G)$. Explain your answer.

2. For the graph G shown in the figure,

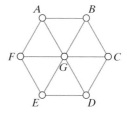

(a) find a 4-coloring of G.

(b) find a 3-coloring of G.

(c) find $\chi(G)$. Explain your answer.

3. For the graph G shown in the figure,

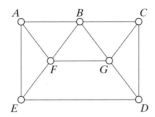

(a) find a 5-coloring of G.

(b) find a 4-coloring of G.

(c) find $\chi(G)$. Explain your answer.

4. For the graph G shown in the figure,

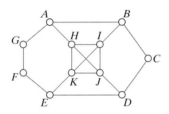

(a) find a 4-coloring of G.

(b) find $\chi(G)$. Explain your answer.

5. For the graph G shown in the figure,

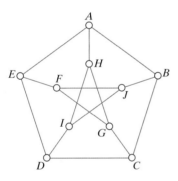

(a) find a 3-coloring of G.

(b) find $\chi(G)$. Explain your answer.

6. For the graph G shown in the figure below,

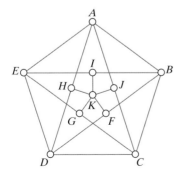

(a) find a 4-coloring of G.

(b) find $\chi(G)$. Explain your answer.

7. Let K_n denote the complete graph on n vertices.

(a) Explain why $\chi(K_n) = n$ (see Example 2.2).

(b) Let G be a graph obtained by removing just one edge from K_n. Find $\chi(G)$. Explain your answer.

8. Explain why if $\chi(G) = 1$, then G consists of just isolated vertices.

9. Let C_n denote the circuit with n vertices (see Example 2.3).

(a) When n is even, $\chi(C_n) = 2$. Describe a 2-coloring of C_n.

(b) When n is odd, $\chi(C_n) = 3$. Describe a 3-coloring of C_n. Explain why a 2-coloring is impossible.

10. Let W_n denote the "wheel" with n vertices, as shown in the figure. Find $\chi(W_n)$. Explain your answer.

(***Hint:*** *You should do Exercise 9 first.*)

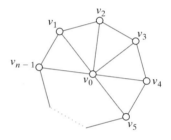

11. If G is a tree (i.e., a connected graph with no circuits), find $\chi(G)$. Explain your answer.

12. The graph shown below is the graph discussed in Example 2.1. Explain why, except for a change of colors, the 3-coloring shown in the figure below is the only possible 3-coloring of the graph.

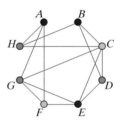

B. Map Coloring

13. Give an example of a map with 10 states that can be colored with just two colors.

14. Give an example of a map with 4 states that cannot be colored using less than four colors.

15. Find a map of South America and then:

(a) Find the dual graph for the map.

(b) Use the greedy algorithm with the vertices ordered by decreasing order of degree to color the dual graph found in (a).

(c) Find the chromatic number of the dual graph found in (a). What does this imply about the coloring of the original map of South America?

16. Color a map of the lower 48 states in the United States using 4 colors. Explain why it is impossible to color the map using less than four colors. (You might find it convenient to do the coloring online at *http://www.sailor.lib.md.us/MD_topics/kid/col_applet/color.html*. If you choose "U.S. Map" from the drop-down menu that shows "Flag," and then click on "Load Image" you will get a blank map of the lower 48 states and a palette of colors for coloring the map.)

C. Miscellaneous

17. A **bipartite graph** is a graph whose vertices can be separated into two sets A and B such that every edge of the graph joins a vertex in A to a vertex in B. A generic bipartite graph is illustrated in the figure.

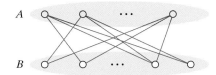

(a) Explain why if G is a bipartite graph, $\chi(G) = 2$.

(b) Explain why if $\chi(G) = 2$, then G must be a bipartite graph.

(c) If G is a bipartite graph, then every circuit in G must have an even number of vertices. True or False? Explain.

18. Suppose G is a graph with n vertices and no loops or multiple edges such that every vertex has degree 3. (These graphs are called *3-regular* graphs.)

(a) Explain why n must be even and $n \geq 4$.

(b) Explain why $\chi(G) \leq 4$.

(c) Describe the 3-regular graphs for which $\chi(G) = 4$. (***Hint***: *What types of 3-regular graphs can fail to meet the conditions of the strong version of Brook's theorem?*)

19. Suppose G is a connected graph with n vertices and no loops or multiple edges such that $n - 1$ vertices have degree 3, and one vertex has degree 2.

(a) Explain why n must be odd and $n \geq 5$.

(b) Explain why $\chi(G) = 3$.
(***Hint***: *Use the strong version of Brook's theorem together with Exercises 8 and 17.*)

20. This exercise refers to the graph in Fig. 2-8(a) (see Example 2.5). Show that if the vertices of the graph are ordered in decreasing order, the greedy algorithm will always give an optimal coloring of the graph.

21. Given a graph G with n vertices, there is a way to order the vertices in the order $v_1, v_2, \ldots, v_n$ so that when the greedy algorithm is applied to the vertices in that order, the resulting coloring of the graph uses $\chi(G)$ colors.
(***Hint***: *Assume a coloring of the graph with $\chi(G)$ colors and work your way backwards to find an ordering of the vertices that produces that particular coloring.*)

22. Let S denote the Sudoku graph discussed in this mini-excursion.

(a) Explain why every vertex of S has degree 20.

(b) Explain why S has 810 edges.
(***Hint***: *See Euler's Theorems in Chapter 5.*)

23. If you haven't done so yet, try the Sudoku puzzle given in Fig. 2-10.

Projects

Scheduling Committee Meetings of the U. S. Senate

This project is a surprising application of the graph coloring techniques we developed in this mini-excursion, and involves an important real-life problem: scheduling meetings for the standing committees of the United States Senate.

The U.S. Senate has 16 different standing committees that meet on a regular basis. The business of these committees represents a very important part of the Senate's work, since legislation typically originates in a standing committee and only if it gets approved there moves on to the full Senate. Scheduling meetings of the 16 standing committees is complicated because many of these committees have members in common, and in such cases the committee meetings cannot be scheduled at the same time. An easy solution

would be to schedule the meetings of the committees all at different time slots that do not overlap, but this would eat up the lion's share of the Senate's schedule, leaving little time for the many other activities that the Senate has to take on. The optimal solution would be to schedule committees that do not have a member in common for the same time slot and committees that do have a member in common for different time slots. This is where graph coloring comes in.

Imagine a graph where the vertices of the graph represent the 16 committees, and two vertices are adjacent if the corresponding committees have one or more members in common. (For convenience, call this graph the *Senate Committees graph*.) If we think of the possible time slots as colors, a k-coloring of the Senate Committees graph gives a way to schedule the committee meetings using k different time slots. In this project you are to find a meetings schedule for the 16 standing committees of the U.S. Senate using the fewest possible number of time slots.

[**Hints**: (1) Find the membership lists for each of the 16 standing committees of the U.S. Senate. You can find the necessary information at *http://www.senate.gov/reference/ resources/pdf/committeelist.pdf.* For the most up to date infor-

mation, go to *http://www.senate.gov* and click on the Committees tab. (2) Create the Senate Committees graph. (You can use a spreadsheet instead of drawing the graph—it is quite a large graph—and get all your work done through the spreadsheet.) (3) List the vertices of the graph in decreasing order of degrees and use the greedy algorithm to color the graph.]

References and Further Readings

1. Delahaye, Jean-Paul, "The Science of Sudoku," *Scientific American*, June 2006, 81–87.

2. *http://www.cut-the-knot.org/Curriculum/Combinatorics/ColorGraph.shtml* (graph coloring applet created by Alex Bogomolny).

3. *http://www.sailor.lib.md.us/MD_topics/kid/col_applet/color.html* (U.S. map coloring applet provided by Sailor, Maryland's Public Information Network).

4. West, Douglas, *Introduction to Graph Theory*. Upper Saddle River, N.J.: Prentice-Hall, Inc., 1996.

5. Wilson, Robin, "The Sudoku Epidemic," *MAA Focus*, January 2006, 5-7.

Solution to Sudoku puzzle

The Time Value of Money

Annuities and Loans

Chapter 10 introduced us to three basic models of population growth (*linear*, *exponential*, and *logistic*), and we saw that these models are applicable to the study of things other than just biological populations. The purpose of this mini-excursion is to discuss in greater detail some of the ideas behind the exponential growth model as they apply specifically to one very important population—your money! You won't find any hot stock tips or get-rich-quick real estate schemes here, but you will gain a better understanding of one of the most important principles of finance—that the value of money is time dependent, and that a dollar in your hands today is worth more than the promise of a dollar tomorrow.

Money has a **present value** and a **future value**. Most of the time, if you give up the right to x dollars today (present value) for a promise of getting the money at some future date, you should expect to get, in return for this sacrifice, something more than x dollars (future value). And of course, the same principle also works in reverse—if you are the one getting the x dollars today (either in cash or in goods), you would expect to have to pay back more than x dollars tomorrow.

The difference between the present value and the future value of money is the price that one party is paying for the *risk* that another party is taking. Every promise of future payment carries some element of risk—the risk that the promise will not be kept, and sometimes the risk that the receiving party may not be around to collect. This is a simple-sounding idea, but quantifying risk in the form of dollars and cents involves many variables and is far from trivial. But that is the business of bankers and insurance folk.

Our task here is considerably less ambitious. In this mini excursion we will explore the relationship between the present and future values of money when dealing with *ordinary annuities* and *installment loans*. The only mathematical tools we will be using are the *general compounding formula* and the *geometric sum formula*, both introduced in Chapter 10. For the reader's benefit, here are the formulas revisited.

General Compounding Formula

If a is invested today, the future value of a after t years is $a \cdot r^{t \cdot k}$ where k is the number of compounding periods per year and $r = 1 + p$ [p is the periodic interest rate expressed as a decimal (annual interest rate divided by k)].

Geometric Sum Formula

$$a + ar + ar^2 + \cdots + ar^{N-1} = \frac{a(r^N - 1)}{r - 1}$$

(The geometric sum formula does not apply when $r = 1$.)

Fixed Annuities

Before defining an annuity, we will illustrate the concept with a couple of examples.

> **EXAMPLE 3.1** The Lottery Winner's Dilemma

People who win a major lottery prize are immediately faced with an important financial decision—take the money in payments spread over an extended period of time (usually 25 years) or take a smaller lump sum payment up front.

Imagine you win a major lottery prize. The choice is $350,000 cash today or $30,000 a year paid over 25 years. Which is a better choice? Tough question, but we will be able to come up with an answer by the end of this mini-excursion. ≪

Example 3.1 illustrates the following classic problem: How does a *present value* of $PV (the $350,000) compare with a future value of N annual payments of $PMT (the $30,000)? As a purely mathematical question the answer depends on PV, N, PMT plus one additional variable—the estimated risk (expressed in the form of an interest rate i) of choosing the future value option. (Note that there are also psychological and sociological aspects to this decision. If you take the lump sum and you have no self-control, you may squander the money in a short time and be miserable later. There is the "mooch factor"—all those friends, relatives, and financial advisors that will want a piece of your action, especially when you take the lump sum option. There is the question of your current financial situation and how much debt you are carrying. And on and on and on. The list is long.)

The next example is a variation on the theme first raised in Example 10.17 in Chapter 10.

> **EXAMPLE 3.2** Setting Up a College Trust Fund: Part 1

A mother decides to set up a college trust fund for her newborn child by making equal monthly payments into the trust fund over a period of 18 years (216 total payments). The trust fund guarantees a fixed annual interest rate of 6% compounded monthly, which equates to a periodic interest rate $p = 0.005$ (one-half of a percent). Mom's original plan was to make monthly payments of $100 at the beginning of each month. In this case, at the end of 18 years the total in the trust fund is $38,929. (This total was computed in Example 10.17, but we will go over the details again in the next example.)

Mom's problem is that the $100 monthly payments are leaving her far short of her goal of a $50,000 trust fund at the end of 18 years. What are the monthly payments she should make to the trust fund to reach her $50,000 target in 18 years? We will answer this question in the next section as well. ⟪

Examples 3.1 and 3.2 illustrate the concept of a **fixed annuity**—a sequence of *equal* payments made or received at the end of *equal* time periods. Annuities (often disguised under different names) are so common in today's financial world that there is a good chance you may be currently involved in one or more annuities and not even realize it. You may be *making* regular deposits to save for an expensive item such as a vacation, a wedding, or college, or making regular payments on a car loan or on your credit card debt (ugh!). You could also be at the receiving end of an annuity, *getting* regular payments from an inheritance, a college trust fund set up on your behalf, or a lottery win.

The word *annuity* comes from the Latin *annua*. Ancient Roman contracts called *annua* were sold to individuals in exchange for lifetime payments made once a year. The United Kingdom started the first group annuity called the State of Tontine in 1693 in order to raise money for war. In this annuity people could buy a share of the Tontine for a fixed sum in return for annual payments for the remainder of that person's life. In the United States annuities began to be offered during the Great Depression and have since become an integral part of the modern financial world that sooner or later we all have to face.

For the remainder of this mini-excursion we will focus on two basic types of annuities: *fixed deferred annuities* and *fixed immediate annuities*. A **fixed deferred annuity** is an annuity in which a series of regular payments are made in order to produce a lump sum at a later date; a **fixed immediate annuity** is an annuity in which a lump sum is paid to generate a series of regular payments later. You can think of a deferred annuity as the process of creating a retirement nest egg (the accumulation phase) and an immediate annuity as the process of taking money out of a retirement nest egg (the payout phase).

Deferred Annuities

To measure how good a deferred annuity is we must look at its future value. The **future value** is the sum of all of the payments plus the interest earned. The college trust fund discussed in Example 3.2 is a classic example of a deferred annuity. Let's look at the example in greater detail.

▶ **EXAMPLE 3.3** Setting Up a College Trust Fund: Part 2

In Example 3.2 we mentioned that if $100 is deposited at the beginning of each month in a trust fund that pays 6% annual interest compounded monthly, the future value of the trust fund after 18 years is $38,929. Let's consider in more detail how this number comes about.

The periodic interest rate is $p = 0.06/12 = 0.005$ (an annual interest rate of 6% compounded monthly). From the general compounding formula we have that the first deposit of \$100, compounded over 216 periods (18 years = 216 months), has a future value of $\$100(1.005)^{216}$. The second deposit of \$100, compounded over 215 periods, has a future value of $\$100(1.005)^{215}$. The third deposit of \$100, compounded over 214 periods, has a future value of $\$100(1.005)^{214}$. And so on. The last deposit of \$100 is compounded over only one period and has a future value of \$100(1.005).

The future value of a deferred annuity (let's call it FV) is the sum of the future values of all the deposits. In this case,

$$FV = \$100(1.005)^{216} + \$100(1.005)^{215} + \cdots + \$100(1.005)^2 + \$100(1.005)$$

The preceding sum looks like a good candidate for the geometric sum formula. To best see how the formula applies, we reverse the order of the terms and rewrite 100(1.005) as 100.5. When we do that we get

$$FV = \$100.5 + \$100.5(1.005) + \cdots + \$100.5(1.005)^{214} + \$100.5(1.005)^{215}$$

Now we can let $a = 100.50$, $r = 1.005$, and $N = 216$. Using the geometric sum formula, we get

$$FV = \$\frac{(100.50)[(1.005)^{216} - 1]}{1.005 - 1} \approx \$38,929$$

Before we go on, let's deconstruct the preceding expression for FV. The first factor in the numerator is the initial term in the sum, in this case $a = \$100.50 = \$100 \cdot (1.005)$. It represents the money in the trust fund at the end of the first month, in this case the initial monthly payment of \$100 plus \$0.50 interest for the month. (Note that when the payments are made at the *end* of the month, then the initial term a of the sum is equal to the monthly payment.) The second factor in the numerator (inside the square brackets) represents the expression $[r^N - 1]$, and the denominator represents the expression $r - 1$ (which happens to equal p).

The mother's original goal was to have the future value of the trust fund be around \$50,000. To reach this goal, the mother can either increase the length of time over which she makes the \$100 monthly payments, or, alternatively, increase the amount of the monthly payments. Let's consider both options.

Option 1: What would happen if the monthly payments to the trust fund were to stay at \$100 but the payments were extended to 19 years? To calculate the future value of this deferred annuity, all we have to do is increase the exponent N in the previous expression for FV from 216 to 228. Under this option the future value of the trust fund (rounded to the nearest dollar) becomes

$$FV = \$\frac{(100.5)[(1.005)^{228} - 1]}{1.005 - 1} \approx \$42,570$$

This future value is still far short of the \$50,000 goal, and extending the payments over more years doesn't make much sense (it is, after all, a trust fund to help with college expenses). This is clearly not the way to go.

Option 2: Suppose the monthly payments are increased to \$150 and are still made at the beginning of each month for 18 years. The only number that changes now in the original expression for FV is the first factor in the numerator, which becomes $150(1.005) = 150.75$. Under this option the future value of the trust fund (rounded to the nearest dollar) becomes

$$FV = \$\frac{(150.75)[(1.005)^{216} - 1]}{1.005 - 1} \approx \$58,393$$

This future value overshoots the $50,000 target by quite a bit, so we need to reconsider the monthly payments. What is the correct monthly payment that ensures that the deferred annuity has a future value of $50,000 after 18 years? Let's temporarily call this monthly payment x. Using exactly the same argument we used previously, we set up the following equation:

$$FV = \$\frac{x(1.005)[(1.005)^{216} - 1]}{1.005 - 1} = \$50,000$$

Solving the preceding equation for x gives

$$x = \$50,000 \cdot \frac{(1.005 - 1)}{(1.005)[(1.005)^{216} - 1]} \approx \$128.44$$

When we look at it the right way, the formula for the future value of an ordinary deferred annuity matches exactly the geometric sum formula.

Future Value of a Fixed Deferred Annuity

$$FV = a \cdot \frac{r^N - 1}{r - 1}$$

To get this perfect match, we let $r = 1 + p$, where p is the periodic interest rate expressed as a percent, N be the number of equal payments made to the annuity, and a be the money in the annuity at the end of the first period. [When the payments of $\$PMT$ are made at the *end* of each period, then $a = PMT$; when the payments are made at the *start* of each period (as in Example 3.3), then $a = PMT \cdot r$.]

The formula for the future value of a fixed deferred annuity can be used to compute the size of the periodic payment necessary to reach a specific future value target over a given number of payments N. All we have to do is solve the future value formula for the unknown a:

$$a = FV \cdot \frac{r - 1}{r^N - 1}$$

If the payments are made at the end of each period, then we have $PMT = a$; if the payments are made at the beginning of each period, we have $PMT = a/r$.

Immediate Annuities

There are other types of immediate annuities that are not loans—lottery winners who choose a lump sum up front rather than yearly payments are an example—but making payments on installment loans is clearly something that most of us can relate to.

The flip side of a deferred annuity is an *immediate annuity*. The classic example of an immediate annuity is an *installment loan*, where we get a sum of money now (the present value of the loan) and pay it off in installments. When the installment payments are fixed and made over regular periods (monthly, bimonthly, etc.), we have a *fixed immediate annuity*.

Once again, the geometric sum formula will be our main mathematical tool. We will start with a very simple example to illustrate the key concept of this section—the *present value* of a loan.

> ### EXAMPLE 3.4 Present Value of a Single Payment Loan

Jackie just turned 16 and wants to buy a car—now! Problem is, she has no money. She does have a $6000 inheritance that she can cash in in two years, when she turns 18. She wants to borrow against that inheritance from her dad and offers to pay the entire loan back in a single payment when she turns 18. Her dad agrees to lend her the present value of the $6000 that she will pay back in two years, and he will charge a token interest rate of 1.5% a year compounded yearly. How much money should dad spot Jackie to buy the car?

From dad's point of view, he is "investing" a sum of a (the loan amount) to receive $6000 in two years at an annual interest rate of $i = 0.015$. Since the interest compounds yearly, $p = i = 0.015$. Using the general compounding formula, we have

$$a(1.015)^2 = \$6000$$

and thus

$$a = \frac{\$6000}{(1.015)^2} \approx \$5824$$

In the preceding example, $5824 is the *present value* (PV) of the $6000 that Jackie's dad will get in two years, based on the very low interest rate of 1.5% compounded annually. The present value would be much less if Jackie's dad was dealing with a total stranger, a reflection of the fact that the interest rate (the variable that quantifies the risk factor) would be considerably higher.

In general, we can compute the present value of FV at a time N periods into the future using the following formula. (As always, p denotes the periodic interest rate expressed as a decimal.)

Present Value of FV at a Time N Periods in the Future

$$PV = \frac{\$FV}{(1 + p)^N}$$

Note: Sometimes it is more convenient to write the formula in the alternative form $PV = \$FV \cdot (1 + p)^{-N}$.

Remember that a positive exponent in a denominator becomes a negative exponent when moved to the numerator, and vice versa.

In Example 3.4 we dealt with a situation where the repayment of the loan is done in a single lump-sum payoff at the end. While this may work for small loans between family members, the typical situation is that we repay an installment loan by making a long (sometimes too long) series of payments. This makes the computation of the present value a bit more involved, but as before, the geometric sum formula will bail us out.

> **EXAMPLE 3.5** Present Value of an Installment Loan

You just landed a really good job and are finally able to buy that sports car you always wanted. Your local dealer is currently advertising a special offer: zero down and 72 monthly payments of $399 a month. The annual interest rate for the financing is 9%, compounded monthly, so that the periodic interest rate is given by $p = 0.09/12 = 0.0075$. As an educated consumer you ask yourself, "How much am I paying for just the car itself, never mind the financing cost?" You can answer this question by finding the present value of the loan.

In an installment loan such as this one, each of the payments is made at a different time in the future and thus has a different present value. *The present value of the loan is the sum of the present values of the individual payments.* In this case the sum of the present values of the 72 individual payments is given by

$$PV = \$399(1.0075)^{-1} + \$399(1.0075)^{-2} + \cdots + \$399(1.0075)^{-72}.$$

The first term in the above sum is the present value of the first payment, the second term is the present value of the second payment, and so on. (For convenience, these present values are expressed using the negative exponent version of the present value formula.)

We can now apply the geometric sum formula to the preceding expression for PV. There are a couple of different ways that this can be done, but the most convenient is to choose a to equal the *smallest* term in the sum, which in this case is $\$399(1.0075)^{-72}$. Thus, $a = \$399(1.0075)^{-72}$, $r = 1.0075$, and $N = 72$. The geometric sum formula combined with a little algebraic manipulation gives:

See Exercise 7.

$$PV = \$399(1.0075)^{-72} \cdot \frac{[(1.0075)^{72} - 1]}{1.0075 - 1} = \$399 \cdot \frac{[1 - (1.0075)^{-72}]}{0.0075} \approx \$22{,}135$$

The bottom line is that you are paying $\$399 \times 72 = \$28{,}728$ to buy a $22,135 car (the present value of the loan), and the $6593 difference is your financing cost (i.e., the amount of interest you end up paying). ≪

Given the periodic payment PMT, the periodic interest p, and the number of payments N, the present value of any ordinary immediate annuity can be obtained from the *present value formula*. The derivation of the formula is a direct generalization of the computations in Example 3.5.

See Exercise 8.

Present Value of a Fixed Immediate Annuity

$$PV = \$PMT \cdot \frac{[1 - (1 + p)^{-N}]}{p}$$

The root of the word *amortization* is the French word *mort*, meaning "dead" or "killed off"—as in "killing off" the loan.

When we solve the above present value formula for the periodic payment variable PMT, we get an extremely useful formula known as the *amortization formula*. The amortization formula allows us to calculate the size of the payments we need to make to pay off an installment loan with a present value of $\$PV$ given a periodic interest rate p and a number of payments N.

Amortization Formula

$$PMT = \$PV \cdot \frac{p}{[1 - (1 + p)^{-N}]}$$

We are finally in a position to make a sensible decision regarding the lottery winnings discussed in Example 3.1.

▶ EXAMPLE 3.6 The Lottery Winner's Dilemma Revisited

Recall the dilemma facing the lottery winner of Example 3.1 (wasn't it you?): Take the present value of the lottery winnings ($350,000 in cash today) or an annuity of $30,000 a year paid over 25 years. To compare these two options we need to choose a reasonable annual rate of return on the present value (think of it as the rate of return you would expect to get if you took the $350,000 and invested it on your own).

Let's start our comparison with the assumption that we can get an annual rate of return of 5%. Using the amortization formula, we can compute the annual payments we would get on $PV = $350,000 when $p = 0.05$ and $N = 25$:

$$PMT = \$350,000 \cdot \frac{0.05}{[1 - (1.05)^{-25}]} \approx \$24,833$$

Clearly, under the assumption of a 5% annual interest rate we are much better off choosing the annuity of $30,000 a year paid over 25 years offered by the lottery folks. What if we assumed a higher annual rate of return—say, for example, 6%? All we have to do is change the corresponding values in the amortization formula:

$$PMT = \$350,000 \cdot \frac{0.06}{[1 - (1.06)^{-25}]} \approx \$27,379$$

We are still better off taking the annuity. It is only when we assume a rate of return of 7% that the $350,000 lump-sum payment option is a better choice than the annuity option:

$$PMT = \$350,000 \cdot \frac{0.07}{[1 - (1.07)^{-25}]} \approx \$30,034$$

A slightly different way to compare the lump-sum option and the annuity option is by computing the present value of 25 annual payments of $30,000 using the present value formula. Let's start once again assuming an annual interest rate of 5% ($p = 0.05$). Then

$$PV = \$30,000 \cdot \frac{[1 - (1.05)^{-25}]}{0.05} \approx \$422,818$$

If we raise the interest rate to 6% the present value drops to $383,501, and when the interest rate is 7% the present value is $349,607.

The bottom line is that choosing the annuity is essentially equivalent to getting about a 7% rate of return on the present value of $350,000. Looking at it as an investment in the future, the annuity option is the safe, conservative choice. There are, however, many nonmathematical reasons why many people choose

See Exercise 16.

the lump-sum payment option over the annuity—control of all the money, the ability to spend as much as we want whenever we want to, and the realistic observation that we may not be around long enough to collect on the annuity. ◀◀

Conclusion

Whether saving money for a retirement or getting a mortgage to buy a house, understanding the time value of money is crucial to making sound financial decisions. In this mini-excursion we saw how the *general compounding formula* and the *geometric sum formula* (both introduced in Chapter 10) can be combined and modified to give two extremely useful new formulas that allow us to compute the *future value of a fixed deferred annuity* and the *present value of a fixed immediate annuity*. An understanding of these ideas can set us free from the tyranny of bankers and loan officers and from the agony and frustration of incomprehensible finance charges

Exercises

A. The Geometric Sum Formula

1. Use the geometric sum formula to compute
$$\$10 + \$10(1.05) + \$10(1.05)^2 + \cdots + \$10(1.05)^{35}.$$

2. Use the geometric sum formula to compute
$$\$500 + \$500(1.01) + \$500(1.01)^2 + \cdots + \$500(1.01)^{59}.$$

3. Use the geometric sum formula to compute
$$\$10 + \$10(1.05)^{-1} + \$10(1.05)^{-2} + \cdots + \$10(1.05)^{-35}.$$

4. Use the geometric sum formula to compute
$$\$500 + \$500(1.01)^{-1} + \$500(1.01)^{-2} + \cdots + \$500(1.01)^{-59}.$$

5. Use the geometric sum formula to compute $\$10(1.05)^{-1} + \$10(1.05)^{-2} + \$10(1.05)^{-3} + \cdots + \$10(1.05)^{-36}$.

6. Use the geometric sum formula to compute
$$\$500(1.01)^{-1} + \$500(1.01)^{-2} + \cdots +$$
$$\$500(1.01)^{-59} + \$500(1.01)^{-60}.$$

7. Justify each of the following statements.

(a) $\$399(1.0075)^{-1} + \$399(1.0075)^{-2} + \cdots +$

$$\$399(1.0075)^{-72} = \$399(1.0075)^{-72} \cdot \frac{[(1.0075)^{72} - 1]}{1.0075 - 1}$$

(b) $\$399(1.0075)^{-72} \cdot \dfrac{[(1.0075)^{72} - 1]}{1.0075 - 1} =$

$$\$399 \cdot \frac{[1 - (1.0075)^{-72}]}{0.0075}$$

8. Justify each of the following statements.

(*Hint:* Try Exercise 7 first.)

(a) $\$PMT(1 + p)^{-1} + \$PMT(1 + p)^{-2} + \cdots +$

$$\$PMT(1 + p)^{-N} = \$PMT(1 + p)^{-N} \cdot \frac{[(1 + p)^N - 1]}{p}$$

(b) $\$PMT(1 + p)^{-N} \cdot \dfrac{[(1 + p)^N - 1]}{p} =$

$$\$PMT \cdot \frac{[1 - (1 + p)^{-N}]}{p}$$

B. Annuities, Investments, and Loans

9. Starting at the age of 25, Markus invests $2000 at the end of each year into an IRA (individual retirement account). If the IRA earns a 7.5% annual rate of return, how much money is in Markus's retirement account when he retires at the age of 65? (Assume he makes 40 annual deposits, but the last deposit does not generate any interest.)

10. Celine deposits $400 at the end of each month into an account that returns 4.5% annual interest (compounded monthly). At the end of three years she wants to take the money in the account and use it for a 20% down payment on a new home. What is the maximum price of a home that Celine will be able to buy?

11. Donald would like to retire with a $1 million nest egg. He plans to put money at the end of each month into an account carning 6% annual interest compounded monthly. If Donald plans to retire in 35 years, how much does he need to sock away each month?

12. Layla plans to send her daughter to Tasmania State University in 12 years. Her goal is to create a college trust fund worth $150,000 in 12 years. If she plans to make weekly deposits into a trust fund earning 4.68% annual interest compounded weekly, how much should her weekly deposits to the trust fund be? (Assume the deposits are made at the start of each week.)

13. Freddy just remembered that he had deposited money in a savings account earning 7% annual interest at the Middletown bank 15 years ago, but he forgot exactly how much. Today he closed the account and the bank gave him a check for $1172.59. How much was Freddy's initial deposit?

14. **Zero coupon bonds.** A zero coupon bond is a bond that is sold now at a discount and will pay its face value at some time in the future when it matures. Suppose a zero coupon bond matures to a value of $10,000 in seven years. If the bond earns 3.5% annual interest, what is the purchase price of the bond?

15. "*CNN founder and Time Warner vice chairman Ted Turner announced Thursday night that he will donate $1 billion over the next decade to United Nations programs. The donation will be made in 10 annual installments of $100 million in Time Warner stock, he said. 'Present day value that's about $600,000, he joked.*" (**Source**: *CNN, September 19, 1997*)

    Find the present value of this immediate annuity to the United Nations assuming an annual interest rate of 15.1%.

16. **(a)** Find the present value of an annuity consisting of 25 annual payments of $30,000 assuming an annual interest rate of 6%. (Assume the payments are made at the end of each year.)

    (b) Find the present value of an annuity consisting of 25 annual payments of $30,000 assuming an annual interest rate of 7%. (Assume the payments are made at the end of each year.)

    (c) Explain why the present value of an annuity goes down as the interest rate goes up.

17. Ned Flounders plans to donate $50 per week to his church for the next 60 years. Assuming an annual interest rate of $5\frac{1}{5}$% compounded weekly, find the present value of this immediate annuity to Ned's church.

18. Michael Dell, founder of Dell computers, was reputed to have a net worth of $16 billion in 2005. If Mr. Dell were to take the $16 billion and set up a 40-year immediate annuity for himself, what would his yearly payment from the annuity be? Assume a 3% annual rate of return on his money.

19. The Simpsons are planning to purchase a new home. To do so, they will need to take out a 30-year home mortgage loan of $160,000 through Middletown bank. Annual interest rates for 30-year mortgages at the Middletown bank are 5.75% compounded monthly.

    (a) Compute the Simpsons' monthly mortgage payment under this loan.

    (b) How much interest will the Simpsons pay over the life of the loan?

20. The Smiths are refinancing their home mortgage to a 15-year loan at 5.25% annual interest compounded monthly. Their outstanding balance on the loan is $95,000.

    (a) Under their current loan, the Smiths' monthly mortgage payment is $1104. How much will the Smiths be saving in their monthly mortgage payments by refinancing? (Round your answer to the nearest dollar.)

    (b) How much interest will the Smiths pay over the life of the new loan?

21. Ken just bought a house. He made a $25,000 down payment and financed the balance with a 20-year home mortgage loan with an interest rate of 5.5% compounded monthly. His monthly mortgage payment is $950. What was the selling price of the house?

C. Miscellaneous

22. You want to purchase a new car. The price of the car is $24,035. The dealer is currently offering a special promotion—you can choose a $4000 rebate or 0% financing for 72 months. Assuming you can get a 72-month car loan from your bank at an annual rate of 6% compounded monthly, which is the better deal—the 0% financing or the $4000 rebate? Justify your answer by showing your calculations.

23. **Perpetuities.** A **perpetuity** is a constant stream of identical payments with no end. The present value of a perpetuity of $C, given an annual interest rate p (expressed as a decimal), is given by the formula

    $$\$PV = \$C \cdot (1 + p)^{-1} + \$C \cdot (1 + p)^{-2} + \$C \cdot (1 + p)^{-3} + \cdots$$

    (a) Use the geometric sum formula to compute the value of the sum

    $$\$C \cdot (1 + p)^{-1} + \$C \cdot (1 + p)^{-2} + \$C \cdot (1 + p)^{-3} + \cdots + \$C \cdot (1 + p)^{-N}$$

    (b) Explain why as N gets larger, the value obtained in (a) gets closer and closer to $C/p. (Assume $0 < p < 1$.)

24. "*Borrow $100,000 with a 6% fixed-rate mortgage and you'll pay nearly $116,000 in interest over 30 years. Put an extra $100 a month into principal payments and you'd pay just $76,000—and be done with mortgage payments nine years earlier.*" (**Source:** *Philadelphia Inquirer finance column by Jeff Brown, November 1, 2005*)

    (a) Verify that the increase in the monthly payment that is needed to pay off the mortgage in 21 years is indeed close to $100 and that roughly $40,000 will be saved in interest.

    (b) How much should the monthly payment be increased in order to pay off the mortgage in 15 years? How much interest is saved in doing so?

    (c) How much should the monthly payment be increased in order to pay off the mortgage in t years ($t < 30$)? Express the answer in terms of t.

25. Sam started his new job as the mathematical consultant for the XYZ Corporation on July 1, 2005. The company retirement plan works as follows: On July 1, 2006 the company deposits $1000 in Sam's retirement account,

and each year thereafter, on July 1 the company deposits the amount deposited the previous year plus an additional 6%. The last deposit is made on July 1, 2035. In addition, the retirement account earns an annual interest rate of 6% compounded monthly. On July 1, 2035 all deposits and interest paid to the account stop. What is the future value of Sam's retirement account?

26. To refinance or not to refinance? When interest rates drop there are opportunities to refinance an existing home mortgage by paying the up-front expenses of refinancing and getting a mortgage at a lower interest rate. Whether it is worth doing so or not is a decision that confounds most homeowners. This exercise illustrates all the details that need to be considered to make such a decision. Suppose that your original mortgage is a 30-year loan for $150,000 at 7.5% annual interest compounded monthly. Three years after taking out the original loan, you have an opportunity to refinance and take out a new loan at a 6% annual interest rate compounded monthly. There is an up-front refinancing cost of $1500 (closing costs) plus 2 points (2% of the new loan).

(a) Calculate the monthly payment on the original mortgage ($150,000 for 30 years at 7.5% annual rate compounded monthly). Call this number PMT_1. each.

(b) Calculate the outstanding balance on the original loan after making 36 monthly payments of $$PMT_1$ each.

(c) Calculate the monthly payments if you take out a new loan on the balance computed in (b) for 27 years at a 6% annual interest rate compounded monthly. Call this number PMT_2.

(d) The monthly saving in your mortgage payment if you refinance is $MS = PMT_1 - PMT_2$. Calculate the present value of the 27 years of monthly savings of MS assuming a 3% annual rate of return. Call this number PV.

(e) Find the present value of refinancing the loan. [The present value of refinancing the loan is given by $PV - C$, where C represents the cost of refinancing (closing costs plus points).]

Projects and Papers

A. Growing Annuities

Over time the fixed payment from an annuity (such as a retirement account) will get you fewer and fewer goods and services. If prices rise 3% a year, items that cost $1000 today will cost over $1300 in 10 years and over $1800 in 20 years. To combat this phenomenon, *growing* annuities in which the payments rise by a fixed percentage, say 3%, each year have been developed. In this project, you are to discuss the mathematics behind growing annuities. (See references 1 and 3.)

B. Annuities Illustrated

"*It is good educational psychology to explain difficult topics by simple diagrams. Why, then, have annuities not been put in some graphic form?*" [**Source**: *J. Donald Watson (see reference 4)*]

Of 20 textbooks that Watson referenced in 1936, not one showed any type of diagram to aid in the understanding of annuities. In this project, you are to prepare a presentation illustrating the concepts discussed in this mini excursion by way of diagrams, charts, and other visual aids. While graphs and charts that show the balance on a loan or what happens to the money in a retirement account over time are useful, diagrams that illustrate the general concepts related to the time value of money are the ultimate goal of this project.

References and Further Readings

1. Kaminsky, Kenneth, *Financial Literacy: Introduction to the Mathematics of Interest, Annuities, and Insurance*. Lankham, MD: University Press of America, 2003.

2. Shapiro, David, and Thomas Streiff, *Annuities*. Chicago, IL: Dearborn Financial Publishing, 2001.

3. Taylor, Richard W., "Future Value of a Growing Annuity: A Note," *Journal of Financial Education*, Fall (1986), 17–21.

4. Watson, J. Donald, "Annuities Illustrated by Diagrams," *The Accounting Review*, Vol. 11, No. 2 (1936), 192–195.

5. Weir, David R., "Tontines, Public Finance, and Revolution in France and England, 1688–1789," *The Journal of Economic History*, Vol. 49, No. 1 (1989), 95–124.

The Mathematics of Managing Risk

Expected Value

Life is full of risks. Some risks we take because we essentially have no choice—for example, when we fly in an airplane, get in a car, or ride in an elevator to the top of a skyscraper. Some risks we take on willingly and just for thrills—that's why some of us surf big waves, rock climb, or sky-dive. A third category of risks are *calculated risks*, risks that we take because there is a tangible payoff—that's why some of us gamble, invest in the stock market, or try to steal second base.

Implicit in the term *calculated risk* is the idea that there is some type of calculation going on. Often the calculus of risk is informal and fuzzy ("I have a gut feeling my number is coming up," "It's a good time to invest in real estate," etc.), but in many situations both the risk and the associated payoffs can be quantified in precise mathematical terms. In these situations the relationship between the risks we take and the payoffs we expect can be measured using an important mathematical tool called the *expected value*. The notion of expected value gives us the ability to make rational decisions—what risks are worth taking and what risks are just plain foolish.

In this mini-excursion we will briefly explore the concept of the *expected value* (or *expectation*) of a random variable and illustrate the concept with several real-life applications to gambling, investing, and risk-taking in general. The main prerequisites for this mini-excursion are covered in Chapter 15, in particular the concepts covered in Sections 15.4 and 15.5.

Weighted Averages

> **EXAMPLE 4.1** Computing Class Scores

Imagine that you are a student in Prof. Goodie's Stat 100 class. The grading for the class is based on two midterms, homework, and a final exam. The breakdown for the scoring is given in the first two rows of Table 4-1. Your scores are given in the last row. You needed to average 90% or above to get an A in the course. Did you?

TABLE 4-1

	Midterm 1	Midterm 2	Homework	Final exam
Weight (percentage of grade)	20%	20%	25%	35%
Possible points	100	100	200	200
Your scores	82	88	182	190

One of your friends claims that your A is in the bag, since $(82 + 88 + 182 + 190)/600 = 0.9033 \cdots \approx 90.33\%$. A second friend says that you are going to miss an A by just one percentage point, and isn't that too bad! Your second friend's calculation is as follows: You got 82% on the first midterm, 88% on the second midterm, 91% (182/200) on the homework, and 95% (190/200) on the final exam. The average of these four percentages is $(82\% + 88\% + 95\% + 91\%)/4 = 89\%$.

Fortunately, you know more math than either one of your friends. The correct computation, you explain to them patiently, requires that we take into account that (i) the scores are based on different scales (100 points, 200 points), and (ii) the weights of the scores (20%, 20%, 25%, 35%) are not all the same. To take care of (i) we express the numerical scores as percentages (82%, 88%, 91%, 95%), and to take care of (ii) we multiply these percentages by their respective weights (20% = 0.20, 25% = 0.25, 35% = 0.35). We then add all of these numbers. Your correct average is

$$0.2 \times 82\% + 0.2 \times 88\% + 0.25 \times 91\% + 0.35 \times 95\% = 90\%$$

How sweet it is—you are getting that A after all!

TABLE 4-2 GPA's at Tasmania State University by Class

Class	Freshman	Sophomore	Junior	Senior
Average GPA	2.75	3.08	2.94	3.15

> ## EXAMPLE 4.2 Average GPAs at Tasmania State University

Table 4-2 shows GPAs at Tasmania State University broken down by class. Our task is to compute the overall GPA of *all undergraduates* at TSU.

The information given in Table 4-2 is not enough to compute the overall GPA because the number of students in each class is not the same. After a little digging we find out that the 15,000 undergraduates at Tasmania State are divided by class as follows: 4800 freshmen, 4200 sophomores, 3300 juniors, and 2700 seniors.

To compute the overall school GPA we need to "weigh" the average GPA for each class with the relative size of that class. The relative size of the freshman class is 4800/15,000 = 0.32, the relative size of the sophomore class is 4200/15,000 = 0.28, and so on (0.22 for the junior class and 0.18 for the senior class). When all is said and done, the overall school GPA is

$$0.32 \times 2.75 + 0.28 \times 3.08 + 0.22 \times 2.94 + 0.18 \times 3.15 = 2.9562.$$

Examples 4.1 and 4.2 illustrate how to average numbers that have different relative values using the notion of a *weighted average*. In Example 4.1 the scores (82%, 88%, 91%, and 95%) were multiplied by their corresponding "weights"

(0.20, 0.20, 0.25, and 0.35) and then added to compute the weighted average. In Example 4.2 the class GPAs (2.75, 3.08, 2.94, and 3.15) were multiplied by the corresponding class "weight" (0.32, 0.28, 0.22, 0.18) to compute the weighted average. Note that in both examples the weights add up to 1, which is not surprising since they represent percentages that must add up to 100%.

Weighted Average

Let X be a variable that takes the values $v_1, v_2, \ldots, v_N$, and let $w_1, w_2, \ldots, w_N$ denote the respective weights for these values, with $w_1 + w_2 + \cdots + w_N = 1$. The weighted average for X is given by

$$w_1 \cdot v_1 + w_2 \cdot v_2 + \cdots + w_N \cdot v_N$$

Expected Values

The idea of a weighted average is particularly useful when the weights represent probabilities.

 EXAMPLE 4.3 To Guess or Not to Guess
(Depends on the Question): Part 1

The SAT is a standardized college entrance exam taken by over a million students each year. In the multiple choice sections of the SAT each question has five possible answers (A, B, C, D, and E). A correct answer is worth 1 point, and, to discourage indiscriminate guessing, an incorrect answer carries a penalty of 1/4 point (i.e. it counts as $-1/4$ points).

Imagine you are taking the SAT and are facing a multiple choice question for which you have no clue as to which of the five choices might be the right answer—they all look equally plausible. You can either play it safe and leave it blank or gamble and take a wild guess. In the latter case, the upside of getting 1 point must be measured against the downside of getting a penalty of 1/4 points. What should you do?

Table 4-3 summarizes the possible options and their respective payoffs (for the purposes of illustration assume that the correct answer is B, but of course, you don't know that when you are taking the test).

TABLE 4-3

Option	Leave Blank	A	B	C	D	E
Payoff	0	-0.25	1	-0.25	-0.25	-0.25

We will now need some basic probability concepts introduced in Chapter 15. When you are randomly guessing the answer to this multiple choice question you are unwittingly conducting a *random experiment* with sample space $S = \{A, B, C, D, E\}$. Since each of the five choices is equally likely to be the correct answer (remember, you are clueless on this one), you assign equal probabilities of $p = 1/5 = 0.2$ to each outcome. This probability assignment, combined with the information in Table 4-3, gives us Table 4-4.

Also note that standardized tests such as the SAT are designed for "key balance" (each of the possible answers occurs with approximately equal frequency).

TABLE 4-4

Outcome	Correct Answer (B)	Incorrect Answer (A, C, D, or E)
Point payoff	1	−0.25
Probability	0.2	0.8

Using Table 4-4 we can compute something analogous to a weighted average for the point payoffs with the *probabilities as weights*. We will call this the *expected payoff*. Here the expected payoff E is

$$E = 0.2 \times 1 + 0.8 \times (-0.25) = 0.2 - 0.2 = 0 \text{ points}$$

The fact that the expected payoff is 0 points implies that this guessing game is a "fair" game—in the long term (if you were to make these kinds of guesses many times) the penalties that you would accrue for wrong guesses are neutralized by the benefits that you would get for your lucky guesses. This knowledge will not impact what happens with an individual question [the possible outcomes are still a +1 (lucky) or a −0.25 (wrong guess)], but it gives you some strategically useful information: on the average, totally random guessing on the multiple choice section of the SAT neither helps nor hurts!   «

EXAMPLE 4.4 To Guess or Not to Guess (Depends on the Question): Part 2

Example 4.3 illustrated what happens when the multiple choice question has you completely stumped—each of the five possible choices (A, B, C, D, or E) looks equally likely to be the right answer. To put it bluntly, you are clueless! At a slightly better place on the ignorance scale is a question for which you can definitely rule out one or two of the possible answers. Under these circumstances we must do a different calculation for the risk/benefits of guessing.

Let's consider first a multiple choice question for which you can safely rule out one of the five choices. For the purposes of illustration let's assume that the correct answer is B, and that you can definitely rule out choice E. Among the other four possible choices (A, B, C, or D) you have no idea which is most likely to be the correct answer, so you are going to randomly guess. In this scenario we assign the same probability (0.25) to each of the four choices, and the guessing game is described in Table 4-5.

TABLE 4-5

Outcome	Correct Answer (B)	Incorrect Answer (A, C, or D)
Point payoff	1	−0.25
Probability	0.25	0.75

Once again, the *expected payoff* E can be computed as a weighted average:

$$E = 0.25 \times 1 + 0.75 \times (-0.25) = 0.25 - 0.1875 = 0.0625.$$

An expected payoff of 0.0625 points is very small—it takes 16 guesses of this type to generate an expected payoff equivalent to 1 correct answer (1 point). At

0.0625 = 1/16.

the same time, the fact that it is a positive expected payoff means that the benefit justifies (barely) the risk.

   A much better situation occurs when you can rule out two of the five possible choices (say, D and E). Now the random experiment of guessing the answer is described in Table 4-6.

TABLE 4-6		
Outcome	Correct Answer (B)	Incorrect Answer (A or C)
Point payoff	1	$-1/4$
Probability	1/3	2/3

The switch from decimals to fractions is to avoid having to round-off 1/3 and 2/3 in our calculations.

In this situation the expected payoff E is given by

$$E = (1/3) \times 1 + (2/3) \times (-1/4) = 1/3 - 1/6 = 1/6$$

   Examples 4.3 and 4.4 illustrate the mathematical reasoning behind a commonly used piece of advice given to SAT-takers: *guess the answer if you can rule out some of the options, otherwise don't bother.*

For a review of *random variables* see Chapter 16.

   The basic idea illustrated in Examples 4.3 and 4.4 is that of the **expected value** (or **expectation**) of a random variable.

> **Expected Value**
>
> Suppose X is a random variable with outcomes $o_1, o_2, o_3, \ldots, o_N$, having probabilities $p_1, p_2, p_3, \ldots, p_N$, respectively. The expected value of X is given by
>
> $$E = p_1 \cdot o_1 + p_2 \cdot o_2 + p_3 \cdot o_3 + \cdots + p_N \cdot o_N$$

Applications of Expected Value

In many real-life situations, we face decisions that can have many different potential consequences—some good, some bad, some neutral. These kinds of decisions are often quite hard to make because there are so many intangibles, but sometimes the decision comes down to a simple question: Is the reward worth the risk? If we can quantify the risks and the rewards, then we can use the concept of *expected value* to help us make the right decision. The classic illustration of this type of situation is provided by "games" where there is money riding on the outcome. This includes not only typical gambling situations (dice games, card games, lotteries, etc.) but also investing in real estate or playing the stock market.

   Playing the stock market is a legalized from of gambling where people make investment decisions (buy? sell? when? what?) instead of rolling the dice or spinning a wheel. Some people have a knack for making good investment decisions and in the long run can do very well—others, just the opposite. A lot has to do with what drives the decision-making process—gut feelings, rumors, and can't miss tips from your cousin Vinny at one end of the spectrum, reliable information combined with sound mathematical principles at the other end. Which approach would you rather trust your money to? (If your answer is the former then you might as well skip the next example.)

▶ EXAMPLE 4.5 To Buy or Not to Buy?

Fibber Pharmaceutical is a new start-up in the pharmaceutical business. Its stock is currently selling for $10. Fibber's future hinges on a new experimental vaccine they believe has great promise for the treatment of the avian flu virus. Before the vaccine can be approved by the FDA for use with the general public it must pass a series of clinical trials known as Phase I, Phase II, and Phase III trials. If the vaccine passes all three trials and is approved by the FDA, shares of Fibber are predicted to jump tenfold to $100 a share. At the other end of the spectrum, the vaccine may turn out to be a complete flop and fail Phase I trials. In this case, Fibber's shares will be worthless. In between these two extremes are two other possibilities: the vaccine will pass Phase I trials but fail Phase II trials or pass Phase I and Phase II trials but fail Phase III trials. In the former case shares of Fibber are expected to drop to $5 a share; in the latter case shares of Fibber are expected to go up to $15 a share.

Table 4-7 summarizes the four possible outcomes of the clinical trials. The last row of Table 4-7 gives the probability of each outcome based on previous experience with similar types of vaccines.

TABLE 4-7

Outcome of trials	Fail Phase I	Fail Phase II	Fail Phase III	FDA Approval
Estimated share price	$0	$5	$15	$100
Probability	0.25	0.45	0.20	0.10

Combining the second and third rows of Table 4-7, we can compute the expected value E of a future share of Fibber Pharmaceutical:

$$E = 0.25 \times \$0 + 0.45 \times \$5 + 0.20 \times \$15 + 0.10 \times \$100 = \$15.25$$

To better understand the meaning of the $15.25 expected value of a $10 share of Fibber Pharmaceutical, imagine playing the following game: For a cost of $10 you get to roll a die. This is no ordinary die—this die has only four faces (labeled I, II, III, and IV), and the probabilities of each face coming up are different (0.25, 0.45, 0.20, and 0.10, respectively). If you Roll a I your payoff is $0 (your money is gone); if you roll a II your payoff is $5 (you are still losing $5); if you roll a III your payoff is $15 (you made $5); and if you roll a IV your payoff is $100 (jackpot!). The beauty of this random experiment is that it can be played over and over, hundreds or even thousands of times. If you do this, sometimes you'll roll a I and poof—your money is gone, a few times you'll roll a IV and make out like a bandit, other times you'll roll a III or a II and make or lose a little money. The key observation is that if you play this game long enough the average payoff is going to be $15.25 per roll of the die, giving you an average profit or gain of $5.25 per roll. In purely mathematical terms, this game is a game well worth playing (but since this book does not condone gambling, this is just a theoretical observation).

Returning to our original question, is Fibber Pharmaceutical a good investment or not? At first glance, an expected value of $15.25 per $10 invested looks like a great risk, but we have to balance this with the several years that it might take to collect on the investment (unlike the die game that pays off right away). For example, assuming four years to complete the clinical trials (sometimes it can take even longer), an investment in Fibber shares has a comparable expected payoff as a safe investment at a fixed annual yield of about 11%. This makes for a good, but hardly spectacular, investment.

See Exercise 20.

◀◀

EXAMPLE **4.6** Raffles and Fundraisers: Part 1

A common event at many fundraisers is to conduct a raffle—another form of legalized gambling. At this particular fundraiser, the raffle tickets are going for $2.00. In this raffle, they will draw one grand prize winner worth $500, four second prize winners worth $50 each, and fifteen third prize winners worth $20 each. Sounds like a pretty good deal for a $2.00 investment, but is it? The answer, of course, depends on how many tickets are sold in this raffle.

Suppose that this is a big event, and they sell 1000 tickets. Table 4-8 shows the four possible outcomes for your raffle ticket (first row), their respective net payoffs after subtracting the $2 cost of the ticket (second row), and their respective probabilities (third row). The last column of the table reflects the fact that if your number is not called, your ticket is worthless and you lost $2.00.

TABLE 4-8				
Outcome	Grand Prize	Second Prize	Third Prize	No Prize
Net gain	$498	$48	$18	−$2.00
Probability	1/1000	4/1000	15/1000	980/1000

From Table 4-8 we can compute the expected value E of the raffle ticket:

$$E = (1/1000) \times \$498 + (4/1000) \times \$48 + (15/1000) \times \$18 + (980/1000) \times \$(-2)$$
$$= \$0.498 + \$0.192 + \$0.27 - \$1.96 = -\$1.00$$

The negative expected value is an indication that this game favors the people running the raffle. We would expect this—it is, after all, a fundraiser. But the computation gives us a precise measure of the extent to which the game favors the raffle: on the average we should expect to lose $1.00 for every $2.00 raffle ticket purchased. If we purchased, say, 100 raffle tickets we would in all likelihood have a few winning tickets and plenty of losing tickets, but who cares about the details—at the end we would expect a net loss of about $100.   ◀◀

EXAMPLE **4.7** Raffles and Fundraisers: Part 2

Most people buy raffle tickets to support a good cause and not necessarily as a rational investment, but it is worthwhile asking, What should be the price of a raffle ticket if we want the raffle to be a "fair game"? (Let's assume we are still discussing the raffle in Example 4.6.) In order to answer this question, we set the price of a raffle ticket to be x, set the expected gain to be $0 (that's what will make it a fair game), and solve the expected value equation for x. Here is how it goes:

$$E = (1/1000) \times \$498 + (4/1000) \times \$48 + (15/1000) \times \$18 + (980/1000) \times \$(-x) = 0$$
$$\$0.498 + \$0.192 + \$0.27 - \$0.98x = 0$$
$$x = \$0.96/0.98 \approx \$0.98$$

◀◀

▶ EXAMPLE 4.8 Chuck-a-Luck

Chuck-a-luck is an old game, played mostly in carnivals and county fairs. We will discuss it here because it involves a slightly more sophisticated calculation of the probabilities of the various outcomes. To play chuck-a-luck you place a bet, say $1.00, on one of the numbers 1 through 6. Say that you bet on the number 4. You then roll three dice (presumably honest). If you roll three 4's you win $3.00; if you roll just two 4's you win $2.00; if you roll just one 4 you win $1.00 (and of course, in all of these cases, you get your original $1.00 back). If you roll no 4's you lose your $1.00. Sounds like a pretty good game, doesn't it?

To compute the expected payoff for chuck-a-luck we will use some of the ideas introduced in Chapter 15, Section 15.5. When we roll three dice, the sample space consists of $6 \times 6 \times 6 = 216$ outcomes. The different events we will consider are shown in the first row of Table 4-9 (the * indicates any number other than a 4). The second row of Table 4-9 shows the size (number of outcomes) of that event, and the third row shows the respective probabilities.

TABLE 4-9

Event	(4,4,4)	(4, 4, *)	(4, *, 4)	(*, 4, 4)	(4, *, *)	(*, 4, *)	(*, *, 4)	(*, *, *)
Size	1	5	5	5	5×5	5×5	5×5	$5 \times 5 \times 5$
Probability	1/216	5/216	5/216	5/216	25/216	25/216	25/216	125/216

See Exercise 19.

Combining columns in Table 4-9, we can deduce the probability of rolling three 4's, two 4's, one 4, or no 4's when we roll three dice. This leads to Table 4-10. The payoffs are based on a $1.00 bet.

TABLE 4-10

Roll	Three 4's	Two 4's	One 4	No 4's
Net gain	$3	$2	$1	−$1
Probability	1/216	15/216	75/216	125/216

The expected value of a $1.00 bet on chuck-a-luck is given by

$$E = (1/216) \times \$3 + (15/216) \times \$2 + (75/216) \times \$1 + (125/216) \times \$(-1)$$
$$= \$(-17/216) \approx \$-0.08$$

Essentially, this means that in the long run, for every $1.00 bet on chuck-a-luck, the player will lose on the average about 8 cents, or 8%. This, of course, represents the bad guys' (i.e. the "house") profit, and in gambling is commonly referred to as the *house margin*, or *house advantage*. ◀◀

The concept of expected value is used by insurance companies to set the price of premiums (the process is called *expectation based pricing*). Our last example is an oversimplification of how the process works, but it illustrates the key ideas behind the setting of life insurance premiums.

> ## EXAMPLE 4.9 Setting Life Insurance Premiums

A life insurance company is offering a $100,000 one-year term life insurance policy to Janice, a 55-year-old nonsmoking female in moderately good health. What should be a reasonable premium for this policy?

For starters, we will let $P denote the *break-even*, or *fair*, premium that the life insurance company should charge Janice if it were not in it to make a profit. This can be done by setting the expected value to be 0. Using mortality tables, the life insurance company can determine that the probability that someone in Janice's demographic group will die within the next year is 1 in 500, or 0.002. The second row of Table 4-11 gives the value of the two possible outcomes to the life insurance company, and the third row gives the respective probabilities.

TABLE 4-11

Outcome	Janice dies	Janice doesn't die
Payoff	$(P - 100,000)	$P
Probability	0.002	0.998

See Exercise 14.

Setting the expected payoff equal to 0 gives $P = 200$. This is the premium the insurance company should charge to break even, but of course, insurance companies are in business to make a profit. A standard gross profit margin in the insurance industry is 20%, which in this case would tack on $40 to the premium. We can conclude that the premium for Janice's policy should be about $240. ◀◀

Exercises

A. Weighted Averages

1. The scoring for Psych 101 is given in the following table. The last row of the table shows Paul's scores. Find Paul's score in the course, expressed as a percent.

	Test 1	Test 2	Test 3	Quizzes	Paper	Final
Weight (percentage of grade)	15%	15%	15%	10%	25%	20%
Possible points	100	100	100	120	100	180
Paul's scores	77	83	91	90	87	144

2. In his record setting victory in the 1997 Masters, Tiger Woods had the following distribution of scores:

Score	2	3	4	5	6
Percentage of holes	1.4%	36.1%	50%	11.1%	1.4%

What was Tiger Woods' average score per hole during the 1997 Masters?

3. At Thomas Jefferson High School, the student body is divided by age as follows: 7% of the students are 14; 22% of the students are 15; 24% of the students are 16; 23% of the students are 17; 19% of the students are 18; the rest of the students are 19. Find the average age of the students at Thomas Jefferson HS.

4. In 2005 the Middletown Zoo averaged 4000 visitors on sunny days, 3000 visitors on cloudy days, 1500 visitors on rainy days, and only 100 visitors on snowy days. The percentage of days of each type in 2005 is given in the following table. Find the average daily attendance at the Middletown Zoo for 2005.

Weather condition	Sunny	Cloudy	Rainy	Snowy
Percentage of days	47%	27%	19%	7%

B. Expected Values

5. Find the expected value of the random variable with outcomes and associated probability distribution shown in the following table.

Outcome	5	10	15
Probability	1/5	2/5	2/5

6. Find the expected value of the random variable with outcomes and associated probability distribution shown in the following table.

Outcome	10	20	30	40
Probability	0.2	0.3	0.4	?

7. A box contains twenty $1 bills, ten $5 bills, five $10 bills, three $20 bills, and one $100 bill. You reach in the box and pull out a bill at random, which you get to keep. Let X represent the value of the bill that you draw.

   (a) Find the probability distribution of X.

   (b) Find the expected value of X.

   (c) How much should you pay for the right to draw from the box if the game is to be a fair game?

8. A basketball player shoots two consecutive free-throws. Each free-throw is worth 1 point and has probability of success $p = 3/4$. Let X denote the number of points scored.

   (a) Find the probability distribution of X.

   (b) Find the expected value of X.

9. A fair coin is tossed three times. Let X denote the number of *Heads* that come up.

   (a) Find the probability distribution of X.

   (b) Find the expected value of X.

10. Suppose you roll a pair of honest dice. If you roll a total of 7 you win $18; if you roll a total of 11 you win $54; if you roll any other total you lose $9. Let X denote the payoff in a single play of this game.

    (a) Find the expected value of a play of this game.

    (b) How much should you pay for the right to roll the dice if the game is to be a fair game?

11. On an American roulette wheel, there are 18 red numbers, 18 black numbers, plus 2 green numbers (0 and 00). If you bet N on red you win N if a red number comes up (i.e. you get $2N$ back—your original bet plus your N profit); if a black or green number come up you lose your N bet.

    (a) Find the expected value of a $1 bet on red.

    (b) Find the expected value of a N bet on red.

12. On an American roulette wheel, there are 38 numbers: $00, 0, 1, 2, \ldots, 36$. If you bet N on any one number— say, for example, on 10—you win $36N$ if 10 comes up (i.e. you get $37N$ back—your original bet plus your $36N$ profit); if any other number comes up you lose your N bet.

    (a) Find the expected value of a $1 bet on 10 (or any other number).

    (b) Find the expected value of a N bet on 10.

13. Suppose you roll a single die. If an odd number (1, 3, or 5) comes up you win the amount of your roll ($1, $3, or $5, respectively). If an even number (2, 4, or 6) comes up, you have to pay the amount of your roll ($2, $4, or $6, respectively).

    (a) Find the expected value of this game.

    (b) Find a way to change the rules of this game so that it is a fair game.

14. This exercise refers to Example 4.9 in this mini-excursion.

    (a) Explain how the fair premium of $P = 200$ is obtained.

    (b) Find the value of a fair premium for a person with probability of death over the next year estimated to be 3 in 1000.

C. Miscellaneous

15. Joe is buying a new plasma TV at Circuit Town. The salesman offers Joe a three-year extended warranty for $80. The salesman tells Joe that 24% of these plasma TVs require repairs within the first three years, and the average cost of a repair is $400. Should Joe buy the extended warranty? Explain your reasoning.

16. Jackie is buying a new MP3 player from Better Buy. The store offers her a two-year extended warranty for $19. Jackie read in a consumer magazine that for this model MP3, 5% require repairs within the first two years at an average cost of $50. Should Jackie buy the extended warranty? Explain your reasoning.

17. The service history of the Prego SUV is as follows: 50% will need no repairs during the first year, 35% will have repair costs of $500 during the first year, 12% will have repair costs of $1500 during the first year, and the remaining SUVs (the real lemons) will have repair costs of $4000 during their first year. Determine the price that the insurance company should charge for a one-year extended warranty on a Prego SUV if it wants to make an average profit of $50 per policy.

18. An insurance company plans to sell a $250,000 one-year term life insurance policy to a 60-year-old male. Of 2.5 million men having similar risk factors, the company estimates that 7500 of them will die in the next year. What is the premium that the insurance company should charge if it would like to make a profit of $50 on each policy?

19. This exercise refers the game of chuck-a-luck discussed in Example 4.8. Explain why, when you roll three dice,

(a) the probability of rolling two 4's plus another number (not a 4) is 15/216.

(b) the probability of rolling one 4 plus two other numbers (not 4's) is 75/216.

(c) the probability of rolling no 4's is 125/216.

20. For this exercise you will need to use the general compounding formula introduced in Chapter 10. Explain why the expected value of an investment in Fibber Pharmaceutical is comparable to an investment with an expected annual yield of about 11%. (Assume it takes four years to complete the clinical trials.)

21. In the California Super Lotto game you choose 5 numbers between 1 and 47 plus a Mega number between 1 and 27. Suppose a winning ticket pays $30 million (assume the prize will not be split among several winners). Find the expected value of a $1 lottery ticket. (Round your answer to the nearest penny.) *Hint*: You may want to review Section 15.3 of Chapter 15.

22. In the California Mega Millions Lottery you choose 5 numbers between 1 and 56 plus a Mega number between 1 and 46. Suppose a winning ticket pays $64 million (assume the prize will not be split among several winners). Find the expected value of a $1 lottery ticket. (Round your answer to the nearest penny.) *Hint*: You may want to review Section 15.3 of Chapter 15.

References and Further Readings

1. Gigerenzer, Gerd, *Calculated Risks*. New York: Simon & Schuster, 2002.

2. Haigh, John, *Taking Chances*. New York: Oxford University Press, 1999.

3. Packel, Edward, *The Mathematics of Games and Gambling*. Washington, DC: Mathematical Association of America, 1981.

4. Weaver, Warren, *Lady Luck: The Theory of Probability*. New York: Dover Publications, Inc., 1963.

Selected Hints

Chapter 1

WALKING

A. Ballots and Preference Schedules

1. **(a)** A majority is more than half.

   (b) To have a plurality is to have more than any other candidate.

   (c) An example of a preference schedule can be found Table 1-1 in the text.

3. **(a)** 8 voters ranked Professor Argand first on their ballot. 3 voters ranked Professor Brandt first.

   (d) The first and last lines of the preference table provide enough data to answer this question.

7. $(0.17)(1500) = 255$ voters prefer L the most; $(0.32)(500) = 480$ voters prefer C the most. The remaining voters (of the 1500 total) prefer M the most.

9. According to the table, 47 voters rank the candidates as follows: B, E, A, C, D.

B. Plurality Method

11. **(a)** Under the plurality method, the candidate with the most first-place votes is declared the winner.

   (b) A, for example, has 108+155 last-place votes. However, be careful to break the tie only by using candidates that have tied with the most first-place votes.

13. **(a)** Suppose that A receives x out of the remaining 30 votes and the competitor with the most votes receives the remaining 30 - x votes. What has to be true of the value of x to guarantee A at least a tie for first?

15. **(a)** A *majority* is more than half (this is very different from a *plurality*).

   (b) The winning candidate has the smallest number of votes possible if the votes are distributed as evenly as possible among the 5 candidates (with one candidate having a slight advantage).

C. Borda Count Method

17. **(a)** Under the Borda count method in this election, a candidate will receive 5 points for each first-place vote, 4 points for each second-place vote, 3 points for each third-place vote, 2 points for each fourth-place vote, and 1 point for each last-place vote.

   (b) The preference schedule of the new election will contain four candidates (A, B, C, and D) since Epstein withdraws from the race. So, the voters will only list these 4 candidates on their preference ballots. The relative rankings on each ballot will remain the same.

   (c) The *independence-of-irrelevant-alternatives* criterion is a measure of fairness. To say it has been violated is to suggest something is askew. Review what the criterion states.

19. Review what the *majority* criterion and *Condorcet* criterion state.

21. **(a)** Column 1 has $0.40 \times 200 = 80$ voters. Determine how many voters correspond to the other columns before tallying Borda points.

23. **(a)** To achieve the maximum number of points possible, the election would need to be a landslide.

(b) If they received the minimum number of points possible, a candidate would be ranked really low on quite a few ballots wouldn't they?

25. (a) A voter gives out 4 points to the first-place candidate and fewer points to each lower ranked candidate.

(b) You would think it would be 110 times as much as the number of points given out by one voter.

(c) Here you need to assume there are 110 voters in the election. Of all the points possible in part (b), *D* will get the number of points that remain.

D.   Plurality-with-Elimination Method

27. (a) Round 1:

Candidate	*A*	*B*	*C*	*D*	*E*
Number of first-place votes	8	3	5	5	0

E is eliminated. The number of first-place votes for the other candidates remains unchanged. Since no candidate has a majority of first-place votes, at least one more round of elimination is needed. In Round 2, candidate *B* is eliminated and their first-place votes transferred to *A*.

33. (b) A Condorcet candidate is one that beats every other candidate in head-to-head competition.

E.   Pairwise Comparisons Method

35. (a) With 5 candidates there are $4 + 3 + 2 + 1 = 10$ pairwise comparisons to consider: *A* vs. *B*, *A* vs. *C*, *A* vs. *D*, *A* vs. *E*, *B* vs. *C*, *B* vs. *D*, etc. Give one point for each victory and ½ a point for each tie.

39. With five candidates, there are a total of 10 pairwise comparisons (and hence 10 points

distributed). Based on this, determine how many times *E* will win.

F.   Ranking Methods

41. (a) Rank the candidates according to how many first-place votes they each have.

(b) Rank the candidates according to how many Borda points they each have.

(c) Rank the candidates in reverse order of elimination.

(d) Rank the candidates according to how many points they receive from the comparisons.

45. From Example 1.5 we know that the winner using the Borda count method is *B* with 106 points. So, *B* is ranked first. Remove *B* from the preference schedule to form a new preference schedule. The winner of the election between candidates *A*, *C*, and *D* will be ranked second. Remove the winner of the election between *A*, *C*, and *D* to form a new preference schedule consisting of the two remaining candidates. The winner of that election will be ranked third and the loser fourth.

47. (a) The winner, using plurality, is *A* with 12 first-place votes. So, *A* is ranked first. Remove *A* from the preference schedule to form a new preference schedule. The winner of the election between candidates *B*, *C*, and *D* using the plurality method will be ranked second. Remove the winner of the election between *B*, *C*, and *D* to form a new preference schedule consisting of the two remaining candidates. The winner of that election (using the plurality method of course) will be ranked third and the loser fourth.

(b) From Exercise 41(b) we know that the winner using the Borda count method is *A*. Wipe all Borda points away and remove *A* from the election to form a new preference schedule. The winner of the election between candidates B, C, and D using the Borda count method will be ranked second. Remove the winner of the election between B, C, and D to form a new preference schedule consisting of the two remaining candidates. The winner of that election will be ranked third and the loser fourth.

(c) From Exercise 41(c) we know that the winner using plurality-with-elimination is C. Remove C from the election to form a new preference schedule among candidates A, B, and D. The winner of the election between candidates A, B, and D using the plurality-with-elimination method will be ranked second.

(d) From Exercise 41(d) we know that the winner using pairwise comparisons is D. Remove D from the preference schedule to form a new preference schedule among candidates A, B, and C. The winner of the election between candidates A, B, and C using pairwise comparisons will be ranked second.

G. Miscellaneous

51. Look at Example 1.14.

53. $501 + 502 + 503 + \ldots + 3220 =$
$\left(1 + 2 + 3 + \ldots + 3220\right) - \left(1 + 2 + 3 + \ldots + 500\right)$

55. (a) Look at Example 1.15.

57. Each column corresponds to a unique ordering of candidates A, B, and C.

JOGGING

61. Suppose the two candidates are A and B and that A gets a first-place votes and B gets b first-place votes and suppose that $a > b$. Then the preference schedule is

Number of voters	a	b
1st choice	A	B
2nd choice	B	A

Use this schedule to determine the winner using each of the four methods discussed in the chapter.

63. Suppose that in a reelection, the only changes in the ballots are in favor of the winning candidate. Could any other candidate pick up extra first-place votes?

67. (a) Suppose a candidate gets v_1 first-place votes, v_2 second-place votes, v_3 third-place votes, ..., v_N Nth-place votes. Find the values of p (Borda point total) and r (Variation 2 total) and add them together.

(b) Suppose candidates C_1 and C_2 receive p_1 and p_2 points respectively using the Borda count as originally described in the chapter and r_1 and r_2 points under the variation described in this exercise. Show that if $p_1 < p_2$, we have $r_1 > r_2$.

69. (a) Each of the 65 voters gives out $25 + 24 + 23$ Borda points to the top three teams on their ballot.

(b) For each team, let x = the number of second-place votes. Knowing the number of first-place votes and Borda count allows one to determine x.

71. By looking at Dwayne Wade's vote totals, it is clear that 1 point is awarded for each third-place vote. You are trying to determine x, the number of points awarded for each second-place vote.

73. (a) With 23 last-place votes, A is eliminated first.

(b) Think of an example in which a candidate has a plurality of first-place votes, but many last-place votes leading to early elimination.

3

(c) Try to eliminate a "middle" candidate by moving the winning candidate up on some ballots. Arrange the numbers so that a different candidate winds up winning.

Chapter 2

WALKING

A. Weighted Voting Systems

1. (a) Inside the square brackets we always list the quota first, followed by a colon and then the respective weights of the individual players.

   (b) The total number of votes is the sum of the weights of the individual players.

   (c) The weight of P_1 is the first number after the colon. The weight of P_2 is the second number listed after the colon.

   (d) It takes 13 votes to pass a motion. What percentage of the total vote is this number?

3. (a) The quota q must be a whole number more than half of the total votes.

   (b) The quota q should not be larger than the total number of votes.

5. (a) To determine the number of votes each player has (and the value of the quota), suppose that P_4 has 1 vote (or one "block" of votes).

7. A player is a dictator if the player's weight is bigger than or equal to the quota. A player that is not a dictator, but that can single-handedly prevent the rest of the players from passing a motion is said to have veto power. A dummy is a player with no power.

At most one player could be a dictator (P_1). Several players (perhaps even all of them) can have veto power. The most likely candidates for veto power are those with the most votes. [Can P_1 prevent a motion from passing? Can P_2? P_3? etc.] The most likely candidates for dummies are those with the fewest votes.

B. Banzhaf Power

11. (b) Winning coalitions are those groups of players that have enough votes to pass a motion (i.e. meet the quota q). Here are all of the coalitions (winning and losing): $\{P_1\}$, $\{P_2\}$, $\{P_3\}$, $\{P_4\}$, $\{P_1, P_2\}$, $\{P_1, P_3\}$, $\{P_1, P_4\}$, $\{P_2, P_3\}$, $\{P_2, P_4\}$, $\{P_3, P_4\}$, $\{P_1, P_2, P_3\}$, $\{P_1, P_2, P_4\}$, $\{P_1, P_3, P_4\}$, $\{P_2, P_3, P_4\}$, $\{P_1, P_2, P_3, P_4\}$

   (c) A critical player is one whose desertion turns a winning coalition into a losing coalition.

   (d) For each winning coalition found in part (b), determine the number of times each player is critical. According to Banzhaf, a player's power is proportional to how often they are critical relative to the other players.

13. (a) The winning coalitions are $\{P_1, P_2\}$, $\{P_1, P_3\}$, and $\{P_1, P_2, P_3\}$.

15. (b) The quota is one more than in (a), so some winning coalitions may now be losing coalitions. For the ones that are still winning, any players that were critical in (a) will still be critical, and there may be additional critical players. A quick check shows that $\{P_1, P_2, P_5\}$, $\{P_1, P_3, P_4\}$, and $\{P_2, P_3, P_4, P_5\}$ are

now losing coalitions (they all have exactly 10 votes).

19. The key to understanding this problem is to understand how much power D has. The other players will (obviously) share the remaining power equally. The winning coalitions have either 2, 3, or 4 players. How often is D critical in each?

C. Shapley-Shubik Power

23. **(a)** There are $3! = 6$ sequential coalitions of the three players. $< P_2, P_3, \underline{P_1} >$ is one of them with the pivotal player underlined. See also Example 2.17.

    (b) P_1 is underlined 4 times; P_2 is underlined 1 time; P_3 is underlined 1 time. Each underline adds to Shapley-Shubik power.

25. There are $4! = 24$ sequential coalitions of the four players. The pivotal player in twelve of these sequential coalitions is underlined below.
$< P_1, \underline{P_2}, P_3, P_4 >, < P_1, \underline{P_2}, P_4, P_3 >,$
$< P_1, P_3, \underline{P_2}, P_4 >, < P_1, P_3, \underline{P_4}, P_2 >,$
$< P_1, P_4, \underline{P_2}, P_3 >, < P_1, P_4, \underline{P_3}, P_2 >,$
$< P_2, \underline{P_1}, P_3, P_4 >, < P_2, \underline{P_1}, P_4, P_3 >,$
$< P_2, P_3, \underline{P_1}, P_4 >, < P_2, P_3, P_4, \underline{P_1} >,$
$< P_2, P_4, \underline{P_1}, P_3 >, < P_2, P_4, P_3, \underline{P_1} >.$

27. **(a)** Try solving this part by inspection.

    (b) Note that P_2 and P_3 are only pivotal if they vote second right after P_1.

    (c) This is very similar to (b).

    (d) One easy way to solve this - How often will P_3 "tip the scales"?

    (e) The second player in any sequential coalition (for this system) is always pivotal.

29. **(c)** How often will P_3 be a pivotal player? Try computing that in your head for a quick solution.

    (e) Notice that every player must vote for a motion in order for it to pass.

33. D will clearly never be pivotal is the first position. D is also never pivotal is the second position since it loses all tie votes. How about when D is in the third or fourth positions?

D. Miscellaneous

35. Look for the $x!$ or $n!$ key. On some calculators, you may need to look under a probability (PRB) menu.

37. **(a)** $10! = 10 \times (9 \times 8 \times \ldots \times 3 \times 2 \times 1)$

    (c) $11! = 11 \times 10 \times (9 \times \ldots \times 3 \times 2 \times 1)$

39. **(a)** $\dfrac{9!+11!}{10!} = \dfrac{9!}{10!} + \dfrac{11!}{10!} = \dfrac{9!}{10 \times 9!} + \dfrac{11 \times 10!}{10!}$

45. **(a)** If the strongest player, P_1, doesn't have veto power, no other player will either.

    (b) Your answer(s) will be larger than the value(s) found part (a), but not so large that P_2, the next strongest player, gains veto power.

    (c) Start by finding values of q for which P_6, the weakest player, has veto power.

    (d) In this case, you want the next strongest coalition $\{P_1, P_2, P_3, P_4, P_5\}$ to be a losing coalition. This will require that the quota be pretty big (but not more than its maximum value of $q = 6 + 5 + 4 + 4 + 3 + 2 = 24$).

JOGGING

47. Since there are no winning two-player coalitions, every player in a three-player coalition is critical. To decide if a given player in a four-player coalition is critical, one need only look for the remaining players to appear as a winning three-player coalition. If the remaining players form a winning coalition, then that given player is not critical. Similarly in the grand (five-player) coalition, one need only look at the winning four-player coalitions to decide which players are critical.

53. **(d)** Suppose that P is never a pivotal member in a sequential coalition. Then, the players that vote before P could form a winning coalition without P.

55. **(a)** There are rules regarding the quota. See Examples 2.2 and 2.3.

    (b) Only one player P_1 could possibly be a dictator. They have at least as many votes as the quota q.

    (c) To have veto power is to have less power than a dictator. If only one player can have veto power, it must be that player with the largest number of votes.

    (d) That is, which values of q result in P_1 and P_2 or all three players having veto power?

57. **(a)** $[24: 14, 8, 6, 4]$ is just $[12: 7, 4, 3, 2]$ with each value multiplied by 2. So the system $[12: 7, 4, 3, 2]$ simply describes a system in which each vote is worth "2 points."

59. **(a)** Count the number of sequential coalitions consisting of the other 5 players. That is, count how many

sequential coalitions can be formed using $N = 5$ players.

(b) If P_1 is fixed in some position, count the number of ways that the other five players can be arranged. That is, count the number of sequential coalitions that can be formed from these other five players.

(c) Note that P_1 is not pivotal as the fourth, third, second, or first player in a sequential coalition.

(d) Since all of the players other than P_1 in the system have same weight, it follows that they will share the remaining power equally.

65. **(a)** For example, the complement of the losing coalition $\{P_1\}$ is the winning coalition $\{P_2, P_3\}$.

(b) One losing coalition (with 2 votes) is $\{P_2, P_3\}$. The complement of this coalition, $\{P_1, P_4\}$, is a winning coalition (with 3 votes).

(c) If P is a dictator, what can be said about any winning coalition? That is, which player is certain to be a part of every winning coalition?

(d) In a decisive voting system, each losing coalition pairs up with a winning coalition (its complement).

67. **(a)** In each nine-member winning coalition, every member is critical. What happens in coalitions consisting of 10 or more members?

(b) At least nine members are needed to form a winning coalition. So, there are $210 + 638 = 848$ winning coalitions. Since every member is critical in each nine-member coalition, the nine-member coalitions yield a total of 210

× 9 = 1890 critical players. How many critical players are there in the coalitions consisting of 10 or more members?

(c) Basically, you are being asked to explain why each permanent member is critical in each of the 848 winning coalitions.

(d) The five permanent members together have 5 × 848/5080 of the power. The remaining power is shared equally.

Chapter 3

WALKING

A. Fair Division Concepts

1. **(a) & (b)** Let C = the value of the chocolate half (in Alex's eyes). Let S = the value of the strawberry half (in Alex's eyes). Then $C = 3S$ and $C + S = \$12$.

   (c) The piece shown is 60/180 of the strawberry half and 40/180 of the chocolate half.

3. **(a), (b) & (c)** Let C = the value of the chocolate part (in Kala's eyes). Let S = the value of the strawberry part (in Kala's eyes). Let V = the value of the vanilla part (in Kala's eyes). Then, $S = 2V$, $C = 3V$, and $C + S + V = \$12$.

5. **(d)** There is only one piece that Ben considers a fair share. He must get that piece in any fair division.

9. **(a)** Let x denote the value of slices s_2 and s_3 to Abe. Then, $2x + 6.50 = 15.00$.

   (c) Let x denote the value (as a percentage of the whole cake) of

slices s_1, s_2 and s_4 to Cory. Then, $x + x + 3x + x = 1$.

(d) Let x denote the value of slices s_2, s_3 and s_4 to Dana.

(e) To start, Cory must receive s_3.

B. The Divider–Chooser Method

13. **(a)** To David, the total amount of pepperoni, sausage, and mushroom must be the same in each piece.

    (b) To Paul, the total amount of anchovies, mushroom, and pepperoni must be the same in each piece.

15. **(a)** See the hint to 13(a).

    (b) See the hint to 13(b).

17. **(a)** Remember that Mo doesn't care which half he winds up with in the end.

    (b) Let P = the value of the pineapple half (in Jamie's eyes). Let O = the value of the orange half (in Jamie's eyes). Then, $O = 4P$ and $O + P = 100\%$.

    (c) You can use common sense on this one. Jamie likes orange better.

C. The Lone-Divider Method

19. **(a)** If Divine is to receive s_1, then what must Chandra receive?

    (b) If Divine is to receive s_2, then what must Chase receive?

25. **(a)** If C_1 is to receive s_2, then what must C_2 receive?

    (b) If C_1 is to receive s_4, then what must C_2 receive? What does that, in turn, force C_3 to receive?

29. (a) The Divider will be the only player that could possibly value each piece equally.

(b) To determine the bids placed, each player's bids should add up to $480,000.

(c) Use the player's bid lists that were found in part (b).

D.   The Lone-Chooser Method

31. (a) Angela sees the left half of the cake as being worth $18. In her second division, she will create three $6 pieces.

(b) Boris sees the right half of the cake as being worth $15. So, in his second division, he will create three $5 pieces. Suppose he cuts a $x°$ wedge out of the strawberry part. Then,

$$\frac{x°}{90°} \times \$9.00 = \$5.00$$

(c) Since Carlos values vanilla twice as much as strawberry, Carlos would want to take the most vanilla that he could from Angela and from Boris.

33. Boris chooses the piece with all of the vanilla because he views it as being worth $12.00 + $6.00 = $18.00. He will cut his piece into portions that are each worth $6.

35. After Arthur makes the first cut, Brian chooses a piece worth (to him) 100% of the cake! Awesome dude.

(a) Basically, Brian's piece is just like one of all the same flavor (to him).

(b) Arthur places all of the value on the orange half of the piece.

(c) Carl is going to choose as much chocolate and vanilla as he can get his hands on.

39. (a) Remember, Karla is a vegetarian so she will simply divide the vegetarian part.

(b) Based on his tastes, Jared will always divide any piece into equally sized portions.

(c) Karla will once again divide whatever vegetarian part she has into equally sized portions.

(d) When she has a choice, Lori prefers meatball subs to equally sized vegetarian subs.

E.   The Last-Diminisher Method

41. (a) Any player that values the piece when it is their turn as worth more than $6.00 will become a diminisher.

(b) Remember, if a player diminishes the piece in round 1, they make it worth exactly 1/5 of the value of the entire cake.

47. (a) The wedge that Arthur claims should be worth ¼ of the value of the entire cake to him. The entire cake to him is the top half.

(d) Brian could stake two different types of claims at the beginning of round 2 – a wedge of chocolate or a wedge of strawberry. Consider each case.

(e) Because of his value system, Damian would want at least half the remaining cake in round 3.

49. (a) Assume, for example, that Lori values the vegetarian part as being worth $3 and the meatball part as being worth $1. Then, she will cut a share worth 1/3 of the entire $4 sub. She will claim some fraction x of the $3 vegetarian part of the sandwich. Thus, $x(\$3) = \frac{1}{3}(\$4)$.

(b) Karla will diminish the *C*-piece cut by Lori because the vegetarian part is even more valuable to her.

F.   The Method of Sealed Bids

51. (a) Ana bids the most on the desk. Further, her bids total $900 so that a fair share to her is $300. Since the value of the items she received is $180, her preliminary cash settlement is $120.

(b) Split the surplus three ways.

57. Suppose that Angelina bid $*x* on the laptop. Then, she values the entire estate at *x*+$2900. Brad, on the other hand, values the entire estate at $4640. Complete the table below to determine the value of *x*.

	Angelina	**Brad**
Fair Share	(*x*+$2900)/2	$232 (
Value of items received	*x*+$300	$278 (
Prelim. cash		-$460
Share of surplus		
Final cash	$355	-$355

G.   The Method of Markers

59. The key is to spot the first first marker and the first second marker.

65. (a) Quintin thinks the total value is $3 \times \$12 + 6 \times \$7 + 6 \times \$4 + 3 \times \$6 = \$120$, so to him a fair share is worth $30. He would place his markers after every $30 worth of comic books.

(b) Look for the first first marker, first second marker, first third marker, and the last third marker.

JOGGING

69. (a) Two divisions are possible.

(b) Four divisions are possible.

(c) Standoff! See Example 3.6.

71. (a) The total of the valuable area is 30,000 m^2. What is the area of *C*? [Is it 1/3 of the valued area?]

(b) Any cut, no matter how whacky, that divides the remaining property in half will work.

(c) A cut parallel to Park Place which divides the parcel in half is illustrated below. Suppose that the cut is made *x* meters from the bottom. Use the fact that $\dfrac{y}{x} = \dfrac{60}{100}$ and that the area of the bottom trapezoid is be 11,000 m^2 to determine the value of *x*.

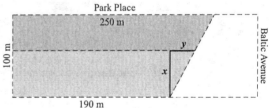

(d) Find *x* so that each shaded area is 11,000 m^2.

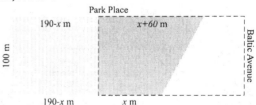

Chapter 4

WALKING

A.   Standard Divisors and Quotas

1. **(a)** The standard divisor represents the number of people represented per seat.

    (b) The standard quota of Apure is the population of Apure divided by the standard divisor. It represents, in a perfectly divisible world, the number of seats in Congress that Apure deserves.

    (c) Upper and lower quotas are found by conventional rounding of standard quotas.

3. **(a)** The "states" in any apportionment problem are the entities that will have "seats" assigned to them according to a share rule.

    (b) The standard divisor is the number of "seats" that each "state" is accorded. See (a).

5. **(a)** If the seats were divisible, each state would receive their standard quota.

    (c) The population of each state is their standard quota × the standard divisor.

7. With 7.43% of the U.S. population, Texas would ideally receive that percentage of the number of seats available in the House of Representatives.

B.   Hamilton's Method

11. Find and sum the lower quotas. Then, allocate any additional seats according to those states whose standard quota have the largest fractional part.

19. **(c)** Notice that for studying an extra 2 minutes (an increase of 3.70%), Bob benefits. However, Peter, who studies an extra 12 minutes (an increase of 4.94%), loses.

21. **(c)** Jim, a new "state," enters the discussion and is given his fair share (6 pieces) of candy.

C.   Jefferson's Method

23. You will need to find a modified divisor (smaller than the standard divisor) for this problem. Luckily, quite a few modified divisors will work. Choose a modified divisor so that the modified lower quotas (the modified quotas rounded down) sum to 160.

25. Any modified divisor between approximately 971 and 975.7 can be used for this problem.

31. Jefferson's method only allows for certain types of quota violations.

D.   Adams' Method

33. You will need to find a modified divisor (larger than the standard divisor) for this problem. This modified divisor will need to be chosen so that the modified upper quotas sum to 160. The modified divisor you choose will need to be less than 51,000.

41. The difference between the number of seats that California would receive under Adams's method and California's standard quota is greater than 1. What fairness criteria does this violate?

E.   Webster's Method

43. Any modified divisor between approximately 49,907 and 50,188 can be used for this problem. Remember too that conventional rounding is the name of the game in Webster's method.

49. Any modified divisor between approximately 0.499% and 0.504% can be used for this problem.

JOGGING

53. **(a)** Both the fractional part and the relative fractional part of the quota for one state should be smaller than for the other state.

    (b) Try making the fractional part of the quota for B bigger than that of A and at the same time the relative fractional part of A bigger than that of B.

    (c) Assume that $f_1 > f_2$, so under Hamilton's method the surplus seat goes to A. Using Lowndes' method, under what condition would the surplus seat go to B?

55. **(a)** In Jefferson's method the modified quotas are larger than the standard quotas. What is the most that rounding these down could do?

    (b) In Adam's method the modified quotas are smaller than the standard quotas.

57. **(a)** Think proportions.

    (b) Use (a) and again think proportions.

59. **(a)** Any modified divisor between approximately 93.8 and 95.2 can be used for Jefferson's method.

    (b) For $D = 100$, the modified quotas are just a bit too small. For $D < 100$, each of the modified quotas will increase (in fact, they will increase more than you might like). For $D > 100$, each of the modified quotas will decrease (which isn't very helpful knowing that $D = 100$ didn't work).

    (c) What choices for modified divisors are there if $D < 100$, $D = 100$, and $D > 100$ do not work?

Chapter 5

WALKING

A.   Graphs: Basic Concepts

1. The degree of a vertex is the number of edges at that vertex. Where there is a loop at a vertex, the loop contributes twice toward the degree.

3. There are an infinite number of ways to do this. One way would be to place all of the vertices on a line in one picture and in the shape of a polygon in the other.

9. Use trial and error but remember that your graph will have symmetry.

11. A path is a sequence of vertices with the property that each vertex in the sequence is adjacent to the next one. Remember that an edge can be part of a path only once. The number of edges in the path is called the length of the path.

    The multiplication rule may come in handy for counting the number of paths between two vertices. For example, there are 3 paths from H to D. Why? There is exactly 1 path from H to A and there are three paths (A, D; A, B, C, D; and A, B, D) from A to D. Putting these facts together gives $1 \times 3 = 3$ paths from H to D.

    Any path from C to F must pass through A and H. The total number of these is the number of paths from C to A times the number of paths from A to H times the number paths from H to F.

13. The only circuit of length 1 is E, E. The only circuit of length 2 is B, C, B. There are circuits between length 3 and 7 in this graph.

B.   Graph Models

15. Represent each student with a vertex.

17. A graph model will use one vertex to represent North Kingsburg and one vertex to represent South Kingsburg.

C.   Euler's Theorems

21. An Euler *circuit* is a circuit that passes through every edge of a graph. Euler's Circuit Theorem can be used to decide if a given graph has an Euler circuit.

    Is the graph connected? Are the vertices all even?

    An Euler *path* is a path that passes through every edge of a graph. Euler's Path Theorem can be used to decide if a given graph has an Euler path.

    Is the graph connected? Are all but exactly two of the vertices even?

D.   Finding Euler Circuits and Euler Paths

27. Fleury's algorithm will always find an Euler circuit for a graph that has one. Typically when a graph has one Euler circuit, it has several.

31. This one is no more complicated than Exercise 27 – just more tedious.

33. In order to have an Euler path, a graph must have exactly two odd vertices. Start by finding these – they will be the beginning and the end of an Euler path. Once again, Fleury's algorithm can typically produce several possible Euler paths.

E.   Unicursal Tracings

37. Euler's Circuit Theorem can be used to decide if a drawing has a closed unicursal tracing (why?). Similarly, Euler's Path Theorem can be used to decide if a drawing has an open unicursal tracing.

    If there is an open or closed unicursal tracing, the process of finding one is exactly

the same as that of finding an Euler circuit or path.

F.   Eulerizations and Semi-Eulerizations

41. The key idea in eulerization is to turn the odd vertices into even vertices by adding "duplicate" (not new) edges in strategic places. Adding as few duplicate edges as possible gives an optimal eulerization.

    A good place to start is to identify (color?) the odd vertices. Then, by connecting pairs of these with sequences of existing edges, the odd vertices will vanish. The difficult part will be adding the fewest number of edges as possible to accomplish this task.

    The odd vertices in (a) and one possible sequence of duplicate edges have been identified for you below.

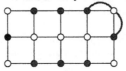

43. The key idea in semi-eulerization is to turn all *but two* odd vertices into even vertices by adding "duplicate" (not new) edges in strategic places. Adding as few duplicate edges as possible gives an optimal semi-eulerization.

G. Miscellaneous

49. Model the tennis court using a graph in which the vertices (at least those of odd degree) are located at "junction points." Then, find a semi-eulerization of that graph.

JOGGING

53. (a) If you were to add a new edge to the graph, how many vertices would increase in degree?

    (b) What would the sum of the degrees of all the vertices be?

55. Try drawing connected regular graphs with $N = 4, 5, 6$, and 7 vertices. For N odd, what prevents you from drawing a graph that doesn't have an Euler circuit?

57. Fleury's algorithm can be paraphrased by "Don't burn your bridges behind you." What is the problem with doing that?

61. (a) Experiment by building graphs for $N = 4$, $N = 5$, and $N = 6$. The strategy that will give you the most money will become clear.

(b) See Example 1.14. Really.

Chapter 6

WALKING

A.   Hamilton Circuits and Hamilton Paths

1. A Hamilton *circuit* in a graph is a circuit in a graph that visits each vertex of the graph once and only once. Similarly, a Hamilton *path* in a graph is a path in a graph that visits each vertex of the graph once and only once.

In this exercise, one Hamilton circuit is A, D, C, E, B, G, F, A.

3. There are eight such Hamilton circuits (four circuits and their mirror images).

First note that edges CD and DE must be a part of every Hamilton circuit and that CE cannot be a part of any Hamilton circuit.

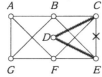

There are 2 possibilities at vertex C: edge BC or edge CF. Edge BC forces edge FE

(otherwise there would be a circuit E, B, C, D, E) and edge CF forces edge EB.

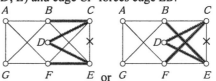

7. Suppose a Hamilton circuit starts by moving to the right (to vertex B). At B there are two choices as to where to go next. Each of these choices completely determines the rest of the circuit.

Starting by moving to the right will force the circuit to end at F. So the mirror image will represent the case of starting by moving left.

9. Any Hamilton path passing through vertex A must contain edge AB. Any path passing through vertex E must contain edge BE. And, any path passing through vertex C must contain edge BC. Consequently, any (Hamilton) path passing through vertices A, C, and E must contain at least three edges meeting at B. This is a problem (why?).

11. The existence of the bridge BC is really the root cause of the lack of Hamilton circuits and certain Hamilton paths in this problem. Why?

13. Don't work too hard on part (c). Remember that Hamilton circuits can be traveled forwards or backwards.

15. There are only two Hamilton paths that start at A and end at C.

B.   Factorials and Complete Graphs

17. The factorial key on your calculator may be hidden under a menu (such as PRB for probability). If the number is large, the calculator will output the number in scientific notation. The two digits on the far right will represent the power of 10 in this notation.

19. The number of distinct Hamilton circuits in K_N is $(N-1)!$.

27. K_N has $N(N-1)/2$ edges.

C.   Brute Force and Nearest Neighbor Algorithms

31. Review the nearest neighborhood circuit constructed in Example 6.7.

33. Use brute-force on (c). There are only 6 possible circuits that make B the first stop after A.

35. See Example 6.8.

D.   Repetitive Nearest-Neighbor Algorithm

37. Essentially, you are applying the nearest-neighbor algorithm five times. This should feel tedious, but not complicated.

The nearest-neighbor circuit found by starting at vertex A has weight 11.0. [This is one-fifth of the work to be done.]

E.   Cheapest-Link Algorithm

43. The first four steps of the cheapest-link algorithm are shown in the following figures.

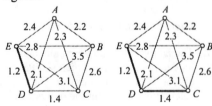

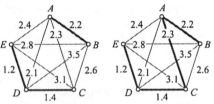

After the fourth step, the next cheapest edge is 2.4. Since this edge makes a circuit, we skip this edge and try the next cheapest edge. The next cheapest edge is 2.6 but that makes three edges come together at vertex C so we skip this edge too.

F.   Miscellaneous

49. There is a lesson to be learned from this exercise. What is it?

51. The weighted graph will represent the six locations using vertices and the weights of the edges between each vertex will represent the distance (the number of blocks a taxi would need to drive) between the vertices.

53. A seating arrangement can be found from a Hamilton circuit.

JOGGING

61. Notice that each of the corner vertices as well as the interior vertex I must be preceded and followed by a boundary vertex.

63. Think of the vertices of the graph as being colored like a checker board (red and black). Why can't you start at a red vertex and end at a red vertex?

67. Is crossing a bridge twice a problem? Why?

Chapter 7

WALKING

A. Trees

1. **(a)** A tree is a network with no circuits. Properties 1-4 will help you decide if a graph is a tree or not.

    (b) Remember that a network is a connected graph.

3. **(a)** How many edges does a tree with 8 vertices have?

    (c) Check out property 2.

5. **(a)** Check out property 1.

    (b) Check out property 2.

9. **(a)** The sum of the degrees of all of the vertices is twice the number of edges.

B. Spanning Trees

11. **(a)** Many answers are possible. Each answer will consist of exactly 4 edges. Two vertices will have degree 1.

    (b) There is only one spanning tree for this network.

13. **(a)** There are three ways to eliminate the only circuit A, B, C, A.

    (b) There are five ways to eliminate the only circuit B, C, D, E, F, B.

15. **(a)** Each spanning tree excludes one of the edges AB, BC, CA and one of the edges DE, EI, IF, FD. The multiplication rule will be useful here. You can review it in chapter 2.

17. **(a)** Each spanning tree excludes one of the edges AB, BC, CA, one of the edges DE, EI, IF, FD, and one of the edges HI, IJ, JH. The multiplication rule is

useful in understanding how to solve this problem.

C. Minimum Spanning Trees and Kruskal's Algorithm

19. **(a)** Add edge EC to the tree first. Then, add edge AD.

    (b) The total weight is $165 + 185 + \ldots$.

23. Kansas City – Tulsa is the cheapest edge. Pierre – Minneapolis is next cheapest. Continue adding edges to the tree without forming a circuit.

25. How many edges will a spanning tree for a graph having 20 vertices have?

D. Steiner Points and Shortest Networks

27. First, determine the measures of each angle in the triangle if possible. If they are all less than 120°, then the network has a Steiner point. If not, the shortest network will be the MST.

29. **(a)** This triangle is isosceles.

    (b) How does the measure of angle ABC compare to that of angle ACB?

31. Since the measure of angle CAB is 120°, there is no Steiner point. Find the lengths of AB and AC since the MST is the shortest network.

33. The Steiner point will create the shortest network. In fact, you can compute the length of that network from the table in this case.

E. Miscellaneous

37. **(a)** The network could have no circuits,

or a circuit of length 3,

or a circuit of length 4,

or ...

(b) There is a pattern in part (a).

39. (a) See property 4.

(b) How many edges will need to be "discarded" in order to form a spanning tree?

(c) Are any edges in the circuit bridges? How about the edges not in the circuit?

43. Try drawing the Steiner tree in. Then, measure every angle in sight.

45. (a) Imagine that the figure is the top half of an equilateral triangle.

(b) By the Pythagorean theorem, $h^2 = s^2 + l^2$. Then, use the result of (a).

47. In a 30°-60°-90° triangle, the side opposite the 30° angle is one half the length of the hypotenuse and the side opposite the 60° angle is $\dfrac{\sqrt{3}}{2}$ times the length of the hypotenuse (See Exercise 45). The measure of angle *BAC* is 60°.

JOGGING

49. (a) Since the circuits C, D, E, C and D, E, I, F, D share a common edge there are two ways to exclude edges to form a spanning tree. If one of the excluded edges is the common edge *DE*, then any one of the other 5 edges *CD, CE, EI, DF*, or *FI* could be excluded to form a spanning tree. If, on the other hand, one of the excluded edges is not the common edge *DE*, then one excluded edge has to be either *CE* or *CD* and the other excluded edge must be either *EI, DF*, or *FI*.

(b) See Example 7.5.

51. (b) Use the fact that the sum of the degrees (which are all the same, namely 1) is also twice the number of edges.

53. (a) See property 3.

(b) Suppose that there were 2 (or more) circuits. What would happen if you removed an edge from one of them (but not the other)?

(c) The maximum redundancy occurs when the degree of each vertex is *N*-1 (i.e. a complete graph).

57. Exhaust all of the possibilities. Using a table will help keep you organized.

61. (a) $\angle BFA$ and $\angle EFG$, for example, are supplementary angles.

(b) Show that the angles in triangle *ABC* are all less than 120°.

(c) Show that a Steiner point *X* could not fall in triangle *ABF*, triangle *BCG*, or triangle *ACE*. That is, show that the angles that *X* would form in those circumstances are not all 120°.

63.

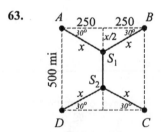

The length of the network is $4x + (500 - x)$

Chapter 8

WALKING

A. Directed Graphs

1. (a) Remember that the arcs are directed so that *AB* is not interpreted the same way as *BA*.

(b) How many arcs come in to each vertex?

3. (a) If *XY* is an arc in the digraph, we say that vertex *X* is *incident to* vertex *Y*.

(b) If *XY* is an arc in the digraph, we say that vertex *Y* is *incident from* vertex *X*.

(e) The arc *YZ* is said to be *adjacent to* the arc *XY* because the starting point of *YZ* is the ending point of *XY*.

5. Many answers are possible. Start by drawing the vertices. Then, draw each arc one at a time.

9. The outdegree of *X* is the number of arcs that have *X* as their starting point (outgoing arcs); the indegree of *X* is the number of arcs that have *X* as their ending point (incoming arcs).

11. Remember that a path must follow in the direction of the arcs. So must a cycle.

15. (a) Is there anyone that everyone respects?

(b) Is there anyone that no one respects?

B. Project Digraphs

17. Try putting tasks that need very little completed before they can start on the left and those that need a lot of other tasks completed before they can start on the right.

19. Perhaps putting the information in a chart such as that given in Exercise 17 would be helpful.

C. Schedules and Priority Lists

23. (a) There are 31 × 3 = 93 processor hours available. How many of these will be used?

(b) How long would it take if there were no idle time?

25. The best case scenario is that the work is divided evenly among all 6 processors.

27. The first processor would start on task *C* (since task *D* is not ready). The second processor would begin on task *A*. Once task *A* is completed, *D* could be started by the second processor. After completing task *C*, the first processor would start task *E*.

31. Remember that the digraph must be followed. So one possible priority list that produces this table is *B,C,A,E,G,D,F* since *D* would be started by the first processor before *G* (since it would not be ready when that point of the list is reached).

33. The general cleaning requires quite a few other tasks to be completed before it is started.

35. This schedule will have quite a bit of idle time for the second work while the first worker is painting. Let's hope you are the second (and not the first) worker in this scenario.

D. The Decreasing-Time Algorithm

37. Decreasing-Time List: *D, C, A, E, B, G, F*. So, the first processor starts with task *C* and the second processor starts with task *A*.

39. Yes, worker 3 has a dream job.

41. (a) The 13 jobs *A–M* are already listed in decreasing order. The precedence relations have little (well, no) effect.

Here is the start of a schedule:

(b) Just a little shuffling at the end of the schedule found in (a) should do the trick. *Opt* = 36.

E.   Critical Paths and Critical-Path Algorithm

45. (a) The critical time for F is $1 + 0 = 1$. The critical time for G is $2 + 0 = 2$. The critical time of D is $12 + 1 = 13$. The critical time for E is $6 + 2 = 8$.

(b) You may be able to eyeball this. It is either a path on the top (*START, A, D, F, END* or *START, B, D, F, END*) or the path on the bottom of the digraph.

(d) What is the best you can do with 2 workers (that must complete 43 hours worth of work)?

47. The critical path list is: $B[46]$, $A[44]$, $E[42]$, $D[39]$, $F[38]$, $I[35]$, $C[34]$, $G[24]$, $K[20]$, $H[10]$, $J[5]$. Here is a start to the schedule:

Time: 0 2 4 6 8 10 12 14 16

P_1	B	E		F
P_2	A	C		Idle
P_3		D		

49. The project digraph, with critical times is:

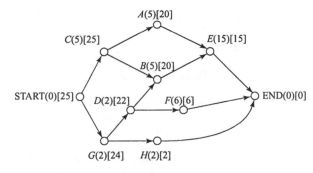

F.   Scheduling with Independent Tasks

51. (a) Remember that the critical-path algorithm is the same as the decreasing-time algorithm when the tasks are independent.

(b) The answer in (a) looks pretty good.

(c) Not much error here.

55. (a) Since all tasks are independent, the critical path list is identical to a decreasing-time list.

(b) *Opt* = 12. Finding the schedule is like putting a small jigsaw puzzle together.

57. (a) This schedule fits together very nicely, thank you.

G.   Miscellaneous

61. (a) This is simply a matter of evaluating the formula for these values of *N*. It may help to convert your answers to percentages.

(b) $\dfrac{M-1}{3M} \le \dfrac{M}{3M}$

JOGGING

71. (a) It may help to think about evenly dividing the work to be done.

(c) $N \times Fin$ is relevant as far as how much you need to *pay* the workers.

Chapter 9

WALKING

A.   **Fibonacci Numbers**

1. (a) The subscript indicates its place in the sequence. Essentially you are being asked to compute the 10th term in the Fibonacci sequence 1,1,2,3,...

     (b) Add 2 *after* finding the value of F_{10}.

     (c) $10 + 2 = 12$

     (d) Divide by 2 *after* finding the value of F_{10}.

     (e) $10/2 = 5$

3. (a) This is a sum of 5 Fibonacci numbers.

     (b) $1 + 2 + 3 + 4 + 5 = 15$

     (c) Order of operations alert: Subscripts take precedence over multiplication.

     (d) $3 \times 4 = 12$

     (e) $F_4 = 3$

5. (a) F_N represents the Nth Fibonacci number. What is being done to that number?

     (b) F_{N+1} represents the Fibonacci number in position $(N + 1)$.

     (c) F_{3N} represents the Fibonacci number in the $3N$th position.

     (d) The $3N+1$ is computed first in evaluating this expression.

7. (a) Fibonacci numbers can be found recursively by $F_N = F_{N-1} + F_{N-2}$.

     (b) Add two Fibonacci numbers to "move forward" in the Fibonacci sequence.

9. You are looking for a way to express the fact that each term of the Fibonacci sequence is equal to the sum of the two preceding terms.

11. (a) 47 is the sum of just two Fibonaccis.

     (b) You'll need three Fibonaccis here. Try using numbers as large as possible.

     (c) Try starting with 144 as one summand. Use the largest Fibonacci number available at each stage of constructing your sum.

     (d) Again, 144 is the largest Fibonacci number less than 210. Try using that in your sum. Think efficiency.

13. (a) The right hand side appears to be the Fibonacci number that follows that last term on the left hand side.

     (b) The subscripts in the fourth equation are 1, 3, 5, 7, 9, and 10. Do you see a pattern? Also note that when the left hand side of the fourth equation ended in F_9, the right hand side was F_{10}.

     (c) The subscript on the right hand side compares closely to the subscript on the last term of the sum.

15. (a) Try the Fibonacci numbers $F_4 = 3$, $F_5 = 5$, $F_6 = 8$, and $F_7 = 13$.

     (b) Denoting the first of these four Fibonacci numbers by F_N, the second of these four Fibonacci numbers is F_{N+1}, the third is F_{N+2}, and the fourth is F_{N+3}.

B.   The Golden Ratio

19. **(a)** The order of operations is critical. First add 1 and $\sqrt{5}$. Next, divide by 2. Then multiply by 21. Finally, add 13.

    (b) Order of operations is again critical. First add 1 and $\sqrt{5}$. Next, divide by 2. Finally, raise the result to the 8th power.

    (c) First add 1 and $\sqrt{5}$. Next, divide by 2. Take the result to the power of 8. Put the result in memory. Next, subtract $\sqrt{5}$ from 1. Then, divide that by 2. Take the resulting number to the power of 8 (as before). Subtract this number from what you have in memory. Finally, divide what you have computed so far by $\sqrt{5}$.

23. **(a)** The golden ratio and the Fibonacci numbers are related in many ways.

    (b) $\phi^9 = 34\phi + 21$

25. The hint also says that $F_N \approx \phi \times F_{N-1}$. Be sure to convert the result of any calculations to scientific notation.

27. **(a)** To find the value of A_7, double the 6th term and add the 5th term. That is,
    $$A_7 = 2A_{7-1} + A_{7-2}$$
    $$= 2A_6 + A_5$$

    (b) That is, estimate the value of $\dfrac{A_7}{A_6}$.

    (c) To find $\dfrac{A_{11}}{A_{10}}$, you will need to find the first 11 terms of the sequence using the given recursive formula.

    (d) Just like ratios of successive Fibonacci numbers settle down to a magical number (in that case the golden ratio), so do ratios of successive terms in the Fibonacci sequence of order 2.

C.   Fibonacci Numbers and Quadratic Equations

29. **(a)** Rewrite this equation as
    $$x^2 - 2x - 1 = 0.$$
    Here $a = 1$, $b = -2$, and $c = -1$. Then,
    $$x = \frac{-(-2) \pm \sqrt{(-2)^2 - 4(1)(-1)}}{2(1)}.$$

31. **(a)** Use trial and error on this one.

    (b) Rewrite the equation:
    $$55x^2 - 34x - 21 = 0$$
    Then $a = 55$, $b = -34$ so that the sum of the two solutions is $-b/a = 34/55$.

33. **(a)** What does it mean for $x=1$ to be a "solution" to an equation?

    (b) Rewrite the equation as
    $$F_N x^2 - F_{N-1}x - F_{N-2} = 0.$$

D.   Similarity

35. **(a)** Since R and R' are similar, each side length of R' is 3 times longer than the corresponding side in R.

    (b) Suppose that R and R' were squares. How many times larger would the area of R' be in that case?

37. **(a)** The ratio of the perimeters must be the same as the ratio of corresponding side lengths.

    (b) For two similar triangles T and T', if the side lengths of T' are x times larger than the corresponding sides in T, then how does the area compare? [Note: The same fact holds true for rectangles and squares.]

39. There are two possible cases that need to be considered. First, we might solve

$$\frac{3}{x} = \frac{5}{8-x}.$$

But, it could also be the case that the side of length 3 does not correspond to the side of length 5, but rather the side of length $8 - x$. In this case, we solve $\dfrac{3}{x} = \dfrac{8-x}{5}$. So more than one solution is possible.

E.   Gnomons

41. So 3 is to 9 as 9 is to $c+3$.

43. 8 is to 12 as $1+8+3$ is to $2+12+x$.

45.

	A	B
20		
	10	x

$10 + x$

47. (a) Angle *CAD* and angle *BAC* are supplementary. To determine the measure of angle *BDC*, use the fact that triangle *BDC* must be similar to triangle *BCA* (in order for triangle *ACD* to be a gnomon). Finally, use the fact that the sum of the measures of the angles in any triangle is 180°.

(b) Use the fact that triangle *DBC* is isosceles.

49. The legs of the two similar triangles must be proportional. So, 3 is to 4 as 9 is to x.

JOGGING

51. The terms in the sequence given appear to have a common factor.

53. (a) $T_1 = 7F_2 + 4F_1$ and
$T_2 = 7F_3 + 4F_2$

(b) Notice that
$$T_{N-1} + T_{N-2} = (7F_N + 4F_{N-1})$$
$$+ (7F_{N-1} + 4F_{N-2})$$

Write these values in terms of F_{N+1} and F_N.

(c) You are being asked to describe T_1, T_2, and the value of T_N in terms of its predecessors.

55. As a similar example, the irrational numbers $\dfrac{5}{3} = 1.\overline{6666}$ and $-\dfrac{2}{3} = -0.\overline{6666}$ sum to an integer 1. So, they have the same decimal expansion.

57. Use the fact that $F_N \phi + F_{N-1} = \phi^N$.

59. If the rectangle is a gnomon to itself, then the rectangle in the figure below must be similar to the original *l* by *s* rectangle.

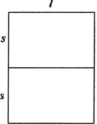

Use this to determine the value of *l/s*.

61. Since the area of the white triangle is 6, the area of the shaded figure must be 48, which makes the area of the new larger similar triangle 6+48=54. Since the ratio of the areas of similar triangles is the square of the ratio of the sides, we have, for example,
$$\frac{3+x}{3} = \sqrt{\frac{54}{6}} = 3.$$

63. The ratio of the area of similar rectangles is the square of the ratio of the sides.

65. (a) Show that triangle *ABD* is isosceles with $AD = AB = 1$. Therefore $AC = x - 1$. Using these facts and the similarity of triangle *ABC* and triangle *BCD* we have

$$\frac{x}{1} = \frac{1}{x-1}.$$

(b) The fact that every triangle in sight is isosceles in this problem should help.

67. (a) Split the regular decagon into ten equally sized triangles all having a vertex at the center. These triangles are each isosceles 36-72-72 triangles. Exercise 65 tells you all you need to know about such triangles to compute the perimeter.

(b) All ten triangles in this part of the problem have side lengths *r* times larger than in the previous part.

Chapter 10

WALKING

A.   Linear Growth and Arithmetic Sequences

1. (a) Each term is 125 larger than the term that preceded it. First, compute $P_1 = P_0 + 125$ and then use that value to find $P_2 = P_1 + 125$.

(b) To find P_1 one would add 125 to $P_0 = 80$. To find P_2, one adds two 125s to $P_0 = 80$. To find P_3, one adds three 125s to $P_0 = 80$. How many 125s are added to $P_0 = 80$ to find P_{100}?

(c) You are being asked to express P_N in terms of *N*. Determine how many 125s are added to $P_0 = 80$ to find P_N.

3. (a) The common difference is what is added at each of the 30 stages of growth.

(b) If $P_0 = 75$ represents the original population, and $P_1 = 80$ represents the population after 1 generation, you are being been asked to find the first value of *N* so that $P_N \geq 1000$. That is, the first value of *N* such that $75 + 5 \times N \geq 1000$.

(c) You are being been asked to find the first value of *N* so that $P_N \geq 1002$.

5. (a) The common difference *d* is added at each of the 10 generations.

(b) The population grew by 30 after 10 generations. Since it grows linearly, it will grow by another 30 after 10 more generations. So, $P_{20} = 68$. Similarly, one can determine P_{30}, P_{40}, and P_{50}.

(c) You are being asked to express P_N in terms of *N*. *N* of the common differences *d* (found in (a)) are added to $P_0 = 8$ to find P_N.

7. (a) Since this is an arithmetic sequence, consecutive terms differ by a common difference *d*. Determine that value. Find the value of A_3 by adding the common difference to A_2.

(b) To get from A_0 to A_1, the common difference *d* is added.

(c) Note that the sequence is decreasing.

9. (a) You are being asked to express P_N in terms of P_{N-1}. The recursive description has the form:
$$P_N = P_{N-1} + d; P_0 = ?$$
Read the equation as "the population of neckties after *N* months (P_N) is the population of neckties after *N*-1 months

(P_{N-1}) plus that number of neckties d bought during that month."

(b) You are being asked to express P_N in terms of N. The explicit description has the form: $P_N = P_0 + N \times d$. Your job is to determine the values of d and P_0.

(c) This is the number of ties Mr. G.Q. has at the end of the 300th month. He starts with $P_0 = 3$ and adds 5 neckties per month for 300 months.

11. (a) This sum is 2 plus how many 5's?

(b) First, determine the type of sequence that the terms of the sum form.

13. (a) Start by finding the common difference d. Then, determine how many of these are added to 12 to arrive at 309.

(b) Determine the type of sequence that appears in this problem. To find the sum, you will need to use the number of terms found in part (a).

15. (a) Since the population grows according to a linear growth model, the terms are that of an arithmetic sequence. Start by determining the value of P_{999}.

(b) $P_{100} + P_{101} + \ldots + P_{999}$
$= (P_0 + P_1 + P_2 + \ldots + P_{999})$
$\quad - (P_0 + P_1 + P_2 + \ldots + P_{99})$

17. (a) Suppose that the number of streetlights at the end of week N is P_N.

(b) Find an explicit expression that represents P_N.

(c) The cost to operate one light for 52 weeks is $52. To operate 137 lights is many times more expensive.

(d) Determine the yearly cost of operating a set of lights installed at the end of week N. Then, add these costs for N between 1 and 52.

B.   Exponential Growth and Geometric Sequences

19. (a) The common ratio r is the value of $\dfrac{P_1}{P_0}$.

(b) P_9 can be found recursively from P_0 (by multiplying by the common ratio r over and over) or explicitly. The latter is the better choice here.

(c) You are being asked to express P_N in terms of r and P_0.

21. (a) In words, the recursive description says that each term in the sequence is 4 times larger than the previous term.

(b) You are being asked to express P_N in terms of N.

(c) Let the population after N generations be modeled by P_N. Find an explicit model for P_N and then solve $P_N \geq 1,000,000$ for N.

23. (a) You are being asked to express P_N in terms of P_{N-1}. But, $P_N = rP_{N-1}$ where $P_0 = 200$. Determine the value of r that one multiplies by each year to account for an *increase* by 50%.

(b) You are being asked to express P_N in terms of N.

(c) You are being asked to approximate the value of P_{10}. $P_{10} = 200 \times r^{10}$ and you need to find the value of the common ratio r.

23

25. (a) In this geometric sequence, each term is found by multiplying the previous term by $r = 2$. So, the first term is $P_0 = 3$ and the second term is $P_1 = 3 \times 2$. You are asked to find the 101st term P_{100}.

(b) You are being asked to express P_N in terms of N.

(c) Since this is the sum of terms in a geometric sequence, you should apply the geometric sum formula.

(d) Note that
$$P_{50} + P_{51} + \ldots + P_{100}$$
$$= (P_0 + P_1 + \ldots + P_{100}) - (P_0 + P_1 + \ldots + P_{49})$$

C.   Financial Applications

27. (a) You are being asked to compute the price of the item after both the coupon and sale are applied. First, try applying the 15% off coupon to the item and then cut that price by 30%. In the second case, mark the item down 30% first and then apply a 15% off coupon to that price.

(b) Find the final price of the item and subtract it from $100 to determine how much money was saved. Then, convert that amount to a percentage.

(c) If the item costs $100 Canadian dollars or 100 rupees, how would your answer to part (b) change?

29. Each year, the account grows by 9%. This is equivalent to multiplying the amount in the account by 1.09 each year. What is in the account if you multiply the original amount by 1.09 at the end of each year for 4 years?

31. Determine how much is in the account on Jan. 1, 2008. Then, use that amount to determine the balance on Jan. 1, 2011.

35. (a) The periodic interest rate is the annual interest rate divided by the number of periods. And, in case you forgot, there are 12 months in a year.

(b) The $5000 grows for 60 periods (called months) at the periodic rate p found in part (a).

(c) Since the interest is compounded monthly, the interest will earn interest. That is, after one year, you will earn more than 12 cents on every dollar invested. In fact, each $1 invested grows to $\$\left(1 + \dfrac{0.12}{12}\right)^{12} \approx \$1.126825$ after one year. What percent return on $1 is this?

41. The periodic interest rate is $p = \dfrac{0.06}{12} = 0.005$. The Jan. 1 deposit will grow for 11 months to $100(1.005)^{11}$. The Feb. 1 deposit will grow for 10 months to $100(1.005)^{10}$. The Mar. 1 deposit will grow for 9 months to $100(1.005)^9$, and so on. Your task is to add the value of each deposit at the beginning of December:
$$100(1.005) + 100(1.005)^2 + \ldots + 100(1.005)^{11}.$$
This is the sum of $N = 11$ terms of a geometric sequence.

43. (a) The amount after $N = 5$ years is $P_5 = \$10,000$. After finding the annual rate p, we solve $10,000 = P_0 \times (1 + p)^5$ for the initial investment P_0.

(b) This is the same as in (a) only the number of periods is $N = 20$ and the periodic rate p is different.

D.   Logistic Growth Model

45. (a) $p_1 = r \times (1 - p_0) \times p_0$

(b) $p_2 = r \times (1 - p_1) \times p_1$

(c) $p_3 = r \times (1 - p_2) \times p_2$; this decimal represents the percent of carrying capacity present in the third generation.

47. (b) Note that the values of p_N are getting very close to 0. This represents the percent of population present in generation N.

E.   Miscellaneous

55. (a) There is a common ratio between terms.

(b) There is a common difference between terms.

(c) There is neither a common ratio nor a common difference between terms.

(d) There is a common ratio between terms. Can you find it?

(e) There is neither a common ratio nor a common difference between terms.

(f) There is a common difference between terms.

(g) There is a common difference (0) between terms. Then again, there is also a common ratio (1) between terms.

57. (a) $P_2 = P_1 + 2P_0$ and $P_3 = P_2 + 2P_1$.

(b) Think about the fact that adding two even numbers always produces an even number.

JOGGING

59. Determine how much $500 would be worth after two years at an annual interest rate of r. Then set that amount equal to 561.80 and solve for the value of r. You will get a quadratic equation in r; solving it will produce two values. You will want to check if they both make sense.

61. (a) Let a be the length of the shortest side of such a triangle. Then the other sides are of length $a + 1d$ and $a + 2d$. By the Pythagorean theorem, it must follow that $a^2 + (a + d)^2 = (a + 2d)^2$. What does say about the value of a?

(b) Suppose, for simplicity, the shortest side of such a triangle is 1. By the Pythagorean theorem, when the sides are in a geometric sequence with common ratio k, it must follow that $1^2 + k^2 = k^4$. What does this say about the value of k? Try letting $x = k^2$ and use the quadratic formula to find the value of x. As always, cast out any solutions that do not make geometric sense.

63. If T was the starting tuition, then the tuition at the end of one year is 110% of T. That is, $(1.10)T$. The tuition at the end of two years is 115% of what it was after one year. That is, after two years, the tuition was $(1.15)(1.10)T = 1.265T$. That's a 26.5% increase over the course of two years.

65. You are being asked to sum $\underbrace{0.01 + 0.02 + 0.04 + 0.08 + \ldots}_{30 \text{ terms}}$

67. If $P_N = a$ is a term of the geometric sequence with common ratio r, then $P_{N+1} = ar$ and $P_{N+2} = ar^2$ are the following two terms. To satisfy the recursive rule, the last term must be the sum of the first two.

69. $\left(a + ar + ar^2 + \ldots + ar^{N-1} \right)(r - 1)$

$= \quad ar + ar^2 + ar^3 + \ldots + ar^{N-1} + ar^N$
$\quad - a - ar - ar^2 - ar^3 - \ldots - ar^{N-1}$

71. The common ratio r should probably be a fraction, wouldn't 'ya think?

73. This would require $p_{N+1} = p_N$. That is, $p_0 = p_1 = 0.8(1 - p_0)p_0$.

75. Knowing the current population and carrying capacity allows you to determine the value of p_0. Then you can get your hands dirty in calculating p_1, p_2, p_3, and p_4.

Chapter 11

WALKING

1. (a) Upon reflection, the image of P with axis l_1 is found by drawing a line through P *perpendicular* to l_1 and finding the point on this line on the opposite side of l_1 which is the same distance from l_1 as the point P.

(c) Since l_3 has slope of 1, the line perpendicular to it has a slope of -1 (the negative reciprocal. That is, the line perpendicular to l_3 will be parallel to l_4.

(d) Since l_4 has slope of -1, the line perpendicular to it has a slope of 1.

3.

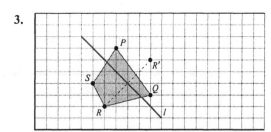

The image of R can be found by drawing a line perpendicular to l and finding the point on the opposite side of this line which is the same distance from l as the point P.

5. (d) A fixed point of a rigid motion is a point that is not moved by that motion.

7. (a) The line segment PP' has a slope of -1 (since moving from P to P' requires one to move down 7 units and right 7 units).

So, the axis of reflection will have a slope of 1 and will pass through the midpoint of segment PP'.

9. A fixed point of a rigid motion is a point that is not moved by that motion. For a reflection, these will always occur on the axis of reflection.

B. Rotations

11. (a) Think of A as the center of a clock in which B is the "9." Where is the "12" of this clock located?

(b) Think of A as the center of a clock in which B is the "9." Where is the "3" of this clock located?

(d) Think of B as the center of a clock in which D is the "1." Where is the "3" located ($60°$ is 1/6, or 2/12, around the clock)?

(e) $120°$ is 1/3, or 4/12, of the distance around a circle.

(g) Think of A as the center of a clock in which I is the "12." Rotating $3690°$ has the same effect as rotating several times all of the way around the clock (plus a remainder). In the remainder lies the answer.

13. (a) Imagine starting at "12" on a clock and rotating clockwise $250°$. How far would you need to rotate the "12" in the other direction in order to end up at the same location? [There are $360°$ in a full circle.]

(b) This is almost two full trips around a circle.

(d) $360° \overline{)3681°}$ $\begin{array}{c} 10 \\ \end{array}$ R $81°$

15. Since BB' and CC' are parallel, the rotocenter will be located at the intersection of the perpendicular bisector of these segments and the line that passes through *BC* (see Figure 11-8(c)).

17. The rotocenter *O* is located at the intersection of the perpendicular bisectors to *PP'* and *SS'*. Since *PP'* has a slope of -1, the its perpendicular bisector will have a slope of 1. Since *SS'* has a slope of -2, its perpendicular bisector will have a slope of ½. Naturally, you should also remember that a perpendicular bisector of a segment passes through the midpoint of that segment.

19. The rotocenter will be located *somewhere* on the perpendicular bisector of *BB'*. To determine where, use a little common sense. Since the rotation is clockwise, the rotocenter must lie on the part of the bisector *below BB'*. Exactly where is specified by the 90° clue.

C.   Translations

21. **(a)** Vector v_1 translates a point 4 units to the right.

    (b) Vector v_2 also translates a point 4 units to the right.

23. Each point moves in the same direction and the same distince under the translation. The location of *E'* unlocks the secret of how each point will move.

D.   Glide Reflections

27. Under a glide reflection, you may glide first and then reflect or reflect first and then glide. It's all up to you.

29. The axis of reflection for this glide reflection passes through the midpoints of *BB'* and *DD'*. Once you determine it, reflect first (to find a figure *A*B*C*D*E** and then glide from there to the final destination called *A'B'C'D'E'*.

33. The axis of reflection cannot be determined by the midpoints of *PP'* and *QQ'* (since they are the same point). So, the axis of reflection is the line perpendicular to *PQ* (see Figure 11-12(c)).

E.   Symmetries of Finite Shapes

35. If you are really stuck on these, try cutting out these shapes with a pair of scissors and fold (reflect) and rotate to discover the symmetries of each.

37. Don't these two look like they are from the same "symmetry family?"

39. Start by determining if there are any reflectional symmetries. If so, it is of type *D*. If not, then it is of type *Z*. The number of rotations the object has determines the value of the subscript.

45. Experiment a bit. Keep your letters "blockish" so that they have the chance at having symmetry (e.g. A has vertical symmetry but A does not).

47. Answers abound. See how creative you can be.

F.   Symmetries of Border Patterns

49. Here is an algorithm you can use: To determine the first symbol (*m* or 1), determine whether the pattern has vertical (*m*idline) symmetry. If so, the first symbol is an *m*, if not, it is a 1.

    To determine the second symbol (*m*, *g*, 2, or 1), determine where the pattern has horizontal (*m*idline), *g*lide reflection, or half-turn (1/2) symmetry. If it has horizontal symmetry, the second symbol is *m*. If it has glide reflectional symmetry, the second symbol is a *g*. If neither of those apply and it has half-turn symmetry, the second symbol is a 2. If none of the above apply, the second symbol will be a 1.

53. Try drawing a object having this symmetry type and the form a border pattern by repeating it in the horizontal direction. That is, develop an example or two of your own.

G.   Miscellaneous

55. Only two of the rigid motions are proper. Only two of the rigid motions have any fixed points. This should narrow your search.

57. (a) The reflection of *P* about l_1 is the point *F*. What is the reflection of *F* about line l_2 ?

    (b) Do what you did in (a) only in the opposite order.

59. (a) Will the left-right and clockwise-counterclockwise orientations on the final figure be the reverse of the original figure? Perhaps an experiment will help.

    (c) This is an improper rigid motion combined with a proper rigid motion (see part a).

61. How many fixed points does the product have?

JOGGING

63. (a) The diagram below shows a reflection from *P* to *P'* and then from *P'* to *P''*.

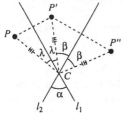

    (b) Relabel the above diagram so that *P''* is *P* and *P* is *P''*.

67. Remember that the only directions for translation symmetries in a border pattern are horizontal.

69. After finding the image in (a), try performing the same rotation and glide reflection again. Think about what types of rigid motions display the resulting property.

71. First, narrow the search down to an improper rigid motion since M is certainly that. What would have to be true about segments PP', RR', and QQ' if this were a reflection?

Chapter 12

WALKING

1. At each stage, each side is replaced by 4 sides having length 1/3 of that being replaced. The length of each side times the number of sides gives the length of the boundary at each stage of construction.

3. At step 1, three triangles are added each having 1/9 the area of the original triangle. That is, the *increase* in area at step 1 is $\dfrac{3 \times 24}{9}$ in^2. At step 2, $3 \times 4 = 12$ triangles are added each having $\dfrac{1}{9^2}$ the area of the original triangle.

5. The snowflake will be 60% larger than the original triangle.

7. At each stage, each side is replaced by 5 sides having length 1/3 of that being replaced. As in Exercise 1, the length of each side times the number of sides gives the length of the boundary at each stage of construction.

9. At step 1, four squares are added each having 1/9 the area of the original square.

That is, the *increase* in area at step 1 is $\frac{4 \times 1}{9}$ in^2. At step 2, $4 \times 5 = 20$ squares are added each having $\frac{1}{9^2}$ the area of the original square. This results in an *increase* in area (from Step 1) of $\frac{20 \times 1}{9^2}$ in^2.

13. At step 1, three triangles are subtracted each having 1/9 the area of the original triangle. That is, the *decrease* in area at step 1 is $\frac{3 \times 81}{9}$ in^2. At step 2, $3 \times 4 = 12$ triangles are subtracted each having $\frac{1}{9^2}$ the area of the original triangle (81 sq. in.).

17. Note that for every square that is added, one like it is taken away.

B.   **The Sierpinski Gasket and Variations**

19. At each step of construction, ¾ of each triangle remains.

21. The area of the Sierpinski gasket is smaller than the area of the gasket formed during any step of construction. Also note that if the area of the original square is 1, then the area at the *N*th stage of construction is $(3/4)^N$.

23. The original triangle is a 6-8-10 triangle (by the Pythagorean theorem). At each stage of construction, every side of each triangle will be ½ of what it was in the previous stage of construction.

27. **(a)** Each removed square will have sides of length 1/3 that of the square from which it was removed. At Step 1, the boundary consists of the previous boundary plus the boundary of a new hole with sides of length 1/3. That is the length of the boundary has increased by $4 \times (1/3)$.

At Step 2, there are 8 new holes, each with sides of length $\frac{1}{3} \times \frac{1}{3} = \frac{1}{9}$. This will *increase* the length of the boundary by $8 \times \frac{4}{9}$.

(b) At step *N+1*, there are 8^N new holes, each with sides of length $\left(\frac{1}{3}\right)^{N+1}$.

29. The area of the gasket at step *N* of the construction will be the product of the number of triangles at that step (6 times more than the previous step) and the area of each such triangle. The length of each side of each triangle in step *N* is 1/3 of the length of a side in step *N*-1.

C.   **The Chaos game and Variations**

33. Start at the first "winning" vertex $P_1 = (32, 0)$. Then, move halfway to the next winning vertex *A* (that is, find the midpoint of *A* and P_1). This will put you at $P_2 = (16, 0)$. Next, move halfway to C, the next winning vertex (i.e. find midpoint of *C* and P_2).

37. Find the new coordinates by picking new *x*-value to be 2/3 of the way from *x*-coordinate of first point to *x*-coordinate of second point and picking new *y*-value to be 2/3 of way from *y*-coordinate of first point to *y*-coordinate of second point.

So, starting at $P_1 = (0, 27)$, move 2/3 of the way to the next winning vertex $B = (27,0)$. $x = 18$ is 2/3 from P_1 to B. $y = 9$ is 2/3 from P_1 to B (the *y*-coordinate is moving down).

D.   **Operations with Complex Numbers**

41. **(a)** $(1+i)^2 + (1+i) = 1 + i + i + i^2 + (1+i)$ and $i^2 = -1$.

(b) $(1-i)^2 + (1-i) = 1 - i - i + i^2 + (1-i)$

45. (a) First, note that $i(1+i) = -1 + i$,
$i^2(1+i) = -1 - i$, $i^3(1+i) = -i + 1$.
Plotting $1+i$ is just like plotting $(1,1)$.
Plotting $-1+i$ is just like plotting $(-1,1)$.

(b) $i(3-2i) = 2 + 3i$, $i^2(3-2i) = -3 + 2i$,
and $i^3(3-2i) = 2 - 3i$.

E.   Mandelbrot Sequences

47. (a) $s_1 = (-2)^2 + (-2)$; $s_2 = (s_1)^2 + (-2)$;
$s_3 = (s_2)^2 + (-2)$; $s_4 = (s_3)^2 + (-2)$.

(b) Don't worry, you should spot a pattern
in (a) that you can apply.

51. (a) $s_1 = \left(\dfrac{1}{2}\right)^2 + \dfrac{1}{2}$; $s_2 = (s_1)^2 + \dfrac{1}{2}$;
$s_3 = (s_2)^2 + \dfrac{1}{2}$.

(b) For example, note that $s_3 > 1$, so we
have $(s_3)^2 > s_3$, and
$s_4 = (s_3)^2 + \dfrac{1}{2} > s_3 + \dfrac{1}{2} > s_3$. What about
the more general case?

(c) Notice from part (a) that $s_2 > 1$.

53. (a) This is an algebra problem at its heart.
Since, $s_{N+1} = (s_N)^2 + s$, you are being
asked to solve $38 = (6)^2 + s$ for s.

(b) A pretty small value of N will work.

JOGGING

55. At each step, one new hole is introduced for
each solid triangle and the number of solid
triangles is tripled. Fill in the following
table to keep your thoughts well-organized.
You will also need to use the geometric sum
formula from chapter 10.

Step	Holes	Solid Triangles
1	1	3
2	$1 + 3$	3^2
3		
4		
N		

57. At the first step, a cube is removed from
each of the six faces and from the center for
a total of 7 cubes removed. At the next step,
each of the 20 remaining cubes has 7 cubes
removed. In the second step, there are
20^2 remaining cubes each of which has 7
cubes removed.

63. Suppose that it is attracted to a number.
Then we have $s_{N+1} = (s_N)^2 + (0.2)$ and we
set $s_{N+1} = s_N$. Substituting, we have
$s_N = (s_N)^2 + 0.2$ so $s_N{}^2 - s_N + 0.2 = 0$.

65. Be bold. Collect several pieces of data by
calculating the first twenty steps in the
sequence. You'll have to get your hands
dirty on this one.

Chapter 13

WALKING

A.   Surveys and Public Opinion Polls

1. (a) The population in this study is the
collection of objects that are under
study.

(b) A sample is a subgroup of the
population from which data is collected.

(c) What is the proportion of red gumballs
in the sample?

(d) There are many different types of sampling methods. They include quota sampling, simple random sampling, convenience sampling, stratified sampling, and taking a census.

3. (a) A parameter is used to describe a population.

(b) A sample statistic is information derived from a sample. It may or may not reflect the actual population parameter.
The difference in these values is attributed as sampling error.

(c) Flip a coin?

5. (a) Find the fraction of the population that makes up the sample.

(b) A sample statistic is information derived from a sample. It may or may not reflect the actual population parameter.

7. Sampling error is the difference between the population parameters (the actual vote) and the sample statistics.

9. (a) While this is a census, it is only one of a subset of the target population.

(b) What fraction of the target population gave their opinion in this survey?

11. (a) The population is the group that we are trying to get information about while the sampling frame consists only of those students that could have been surveyed using this method.

(b) The basic question is whether or not the sample chosen is representative of the population.

17. (a) Who is the survey trying to gain information about?

19. (a) Do you walk about your downtown in a random pattern? Do you think others do?

(b) Assume that people who live or work downtown are much more likely to answer yes than people in other parts of town.

(d) That is, was an attempt made to use quotas to get a representative cross section of the population?

21. (a) Does every member of the population have a chance of being in the sample using this survey technique?

23. (a) In simple random sampling, any two members of the population have as much chance of both being in the sample as any other two.

(b) The students sampled appear to be a reasonably fair cross section of all TSU undergraduates that would attempt to enroll in Math 101.

25. (a) The trees are broken into three different varieties and then a random sample is taken from each.

(b) When selecting 300 trees of variety A, the grower does not select them at random.

B. The Capture-Recapture Method

29. $\dfrac{n_1}{N} = \dfrac{n_2}{k}$

31. To estimate the number of quarters, disregard the nickels and dimes—they are irrelevant.

C. Clinical Studies

39. (a) The target population are those people that could benefit from the treatment. Who are they in this study?

(b) The sampling frame, in this case, is only a small portion of the target population. Be more specific in your response.

(c) Who would the sample likely under represent?

45. (a) There was one group receiving the sham-surgery and two groups receiving a treatment.

(b) The treatments were specific types of knee surgery. Name them.

(c) A randomized study is one in which the patients are assigned to a treatment group or the control group at random.

(d) The question here boils down to who was it that knew the treatment being received by a given patient. Did the patient know? Did the doctors?

47. Was the professor producing a cause and observing the effect? If it was controlled, what was the control group? Were the subjects that received the treatment chosen at random? Was a placebo used?

53. (b) All that is needed for an experiment is for members of the population to receive a treatment. It is a controlled placebo experiment if there is a control group that does not receive the treatment, but instead received a placebo. It is randomized if the participants are randomly divided into treatment and control groups. A study is double blind if neither the participants nor the doctors involved in the clinical trial know who is in each group.

D.   Miscellaneous

55. (a) Was a treatment imposed on a subset of the population?

(b) What group of people is Spurlock trying to make a point about?

(c) What group of people received the treatment?

(d) There are plenty. Almost nothing in the study is well-done.

59. (a) This statement refers to the entire population of students taking the SAT math test.

(b) Only some new automobiles are crash tested.

(c) Not all of Mr. Johnson's blood was used for the test.

JOGGING

65. (a) In the real world, cost and reliability are important factors when making any decision.

(b) People with unlisted phone numbers make up a much higher percentage of the population in a large city such as New York. Also, what do you think is true about people with unlisted phone numbers?

67. Fridays, Saturdays, and Sundays make up 3/7 or about 43% of the week, but people put many more miles on their cars during the work week.

Chapter 14

WALKING

A.   Frequency Tables, Bar Graphs, and Pie Charts

1. A frequency table will have the following format.

Score	10	50	...
Frequency	1	3	

3. (b) The bar graph will have five bars.

5. (a) 24 children scored a 0; 16 children scored a 1

   (b) Determine what fraction of children scored a 0.

7. There are only 9 frequencies to calculate.

9. (b) The central angle of the "very close" slice can be found from the frequency table. 7 of the 25 students fall in this category so that the central angle measures $\frac{7}{25} \times 360° = 100.8°$.

11. (a) two students scored a 3, five students scored a 4, etc...

13. The sum of all of the central angles is 360 degrees.

17. Changing the scale of the vertical axis can really change how one interprets the graph.

B. Histograms

19. (b) 16 oz. = 1 lb.
Also, remember that values that fall exactly on the boundary between two class intervals belong to the class interval to the left (due to the "or equal to" phrase).

   (c) Remember that a histogram is slightly different than a bar graph.

C. Averages and Medians

23. (a) There isn't really a shortcut here – just grind it out.

   (b) First, sort the data. Then, find the locator of the 50th percentile. Based on that, determine the value of the median.

(c) There are shortcuts to finding the average and the median of this new data set. Try to find them.

27. (a) See chapter 1 or chapter 10 for more on how to easily compute the sum $1 + 2 + 3 + ... + 98 + 99$. (Note: This is the sum of terms in an arithmetic sequence.)

   (b) The locator for the median is $L = (0.50)(99) = 49.5$. The locator is not a whole number.

29. (a) The data set consists of $24 + 16 + 20 + ... + 3$ values. 24 of them are 0, 16 are 1, 20 are 2, etc.

   (b) The locator for the median is $L = (0.50)(80) = 40$. So, the median is the average of the 40th and 41st scores. Use the table to determine the 40th and 41st scores (in the order data set).

31. (a) The number of scores is not important. If one likes, they can assume that there were 100 scores.

   (b) Half of the scores are at or below what score?

D. Percentiles and Quartiles

33. (a) The locator of the 25th percentile is $L = (0.25)(9) = 2.25$. So, the first quartile is the 3rd number in the ordered list.

   (b) The locator of the 75th percentile is $L = (0.75)(9) = 6.75$.

   (c) The locator of the 25th percentile of this new data set is $L = (0.25)(10) = 2.5$. The first quartile is still the 3rd number in the ordered list.

37. (a) The Cleansburg Fire Department consists of $N = 2 + 7 + 6 + ...$ firemen. The locator of the first quartile is thus given by $L = (0.25)(N)$.

E.   Box Plots and Five-Number Summaries

41. (a) The five number summary consists of these five numbers: minimum, first quartile, median, third quartile, maximum.

   (b) First, draw and label an appropriate axis. One such grid is shown below.

45. (a) One of the vertical line segments in the box plot will tell you this information.

   (c) Two different vertical line segments in these box plots should explain why this happens.

F.   Ranges and Interquartile Ranges

47. (a) The range is a single number. Your (common sense) answer should be too.

   (b) The IQR is also one number.

51. (a) Note that $1.5 \times IQR = 1.5 \times 3 = 4.5$. So any number more than this amount to the *right* of a box in a box plot is an outlier.

   (b) Any number more than $1.5 \times IQR$ to the *left* of the box in a box plot is an outlier.

G.   Standard Deviations

55. (a) Just how much does this data set deviate from the mean? Do this by inspection if you can.

   (b) The average of the data set is 5. So, the following table is useful. The final step will be to take the square root of the average of the squared deviations.

x	$x - 5$	$(x - 5)^2$
0	−5	25
5	0	0
5	0	0
10	5	25
		50

57. There is a lesson to be learned in doing each of these parts. Standard deviation measures the spread of a data set. Some of these data sets are spread out in the exact same fashion. For part (a), fill out the table shown below:

Data	Deviation from mean	Squared Deviations from mean
0	−4.5	20.25
1	−3.5	12.25
2	−2.5	6.25
3	−1.5	2.25
4	−0.5	0.25
5	0.5	0.25
…	…	…

H.   Miscellaneous

59. One easy way to remember the mode: MOde = MOst common.

JOGGING

65. Let x = Mike's score on the next exam. Find the value of x.

67. Ramon must have done much more poorly than Josh on one of the exams.

71. (a) Each value in a five-number summary is a measure of location. What do you think happens to a measure of location if the same number is added to every value in the data set?

   (b) The "location" of each data point is shifted to the right 10%.

77. Be sure to consider two cases. The first case should be when the data set consists of an odd number of data values.

79. Yes, each of these calculations takes a bit of time and is somewhat messy. The process is tedious but not hard. Be sure to be very careful at every turn.

Chapter 15

WALKING

A.   Random Experiments and Sample Spaces

1. (a) The sample space of a random experiment is the set of all its possible outcomes. One outcome, for example, might be denoted HTT (heads on the first toss, tails on the second toss, heads on the third toss).

(b) The coin may never land heads. Then again, it could always land heads. The sample space will contain each possible number of heads that could be tossed.

(c) A free throw is either successful or not. It is pretty black and white. How many free throws could be made in three tosses?

3. (a) $\{ABCD, ABDC, ACBD, ...\}$ (list 'em all)

(b) There are 4 choices as to which name is drawn first, 3 choices for the next name, etc.

5. A typical outcome is a string of 10 letters each of which can be either an *H* or a *T*. Be as descriptive as you can. There are a lot of outcomes - 2 possible outcomes at each of 10 stages of the experiment.

B.   The Multiplication Rule.

9. (a) You could think of constructing a license plate as a seven stage process. There are 9 possibilities in the first stage, 26 in the second, 26 in the third, etc. [Also, remember that you have 10 digits...they are called fingers...you could label them as 0-9.]

(b) Seems like only one choice at the first and last (seventh) stage of constructing such a plate.

(c) There are 9 possibilities in the first stage, 26 in the second, 25 in the third (no repeats), etc.

11. (a) A password can be typed in 5 stages. The first keystroke could be any of $26 + 26 = 52$ possibilities. Repeats are allowed here.

(b) There is only one possible value for stage one of typing such a password.

(c) There are $25 + 25 = 50$ possibilities for the first stage of building such a password.

(d) Each of the first four stages of construction now has only 50 possibilities.

13. (a) 8 choices for the label of the book at the far left, 7 choices for the label of the book to its right, 6 choices for the label of the book to its right, etc.

(b) How many ways for the books to be in order?

15. (a) Select the committee in three stages. First, select a President (35 choices), then, select a VP (34 choices), etc.

(b) Select the committee in three stages. First, select a President, then, select a VP from the remaining members, etc.

(c) How many all-girl committees are possible?
How many all-boy committees are possible?

17. (a) The only restrictions seem to be that the first digit is not zero and that the last digit is even.

(b) Do you remember your rules of divisibility?

(c) List several numbers that are divisible by 25 if you need to in order to spot a pattern.

C. **Permutations and Combinations**

19. (a) $_nP_r = \dfrac{n!}{(n-r)!}$

(b) $_nC_r = \dfrac{n!}{r!(n-r)!}$

(c) $_{10}P_3 = \dfrac{10!}{7!} = \dfrac{10 \times 9 \times 8 \times \cdots \times 2 \times 1}{7 \times 6 \times 5 \times \cdots \times 2 \times 1}$.

Can you cancel simplify this expression before performing any multiplication?

23. $_{20}C_2 = \dfrac{20!}{18!2!} = {}_{20}C_{18}$

Note also that if the factors of 20! and 18! are written out there is enough simplification to finish the calculation by hand.

25. So apparently the Nth row in Pascal's triangle has a sum that follows a very simply pattern.

27. $_{150}P_{51} = \dfrac{150 \times 149 \times 148 \times \cdots \times 3 \times 2 \times 1}{99 \times 98 \times 97 \times \cdots \times 3 \times 2 \times 1}$

Write out the value of $_{150}P_{50}$ in the same way to determine how many times larger than $_{150}P_{50}$ this is.

29. Sometimes the $_nP_r$ and $_nC_r$ keys can be found under a menu (such as MATH or PRB) on the calculator. Also, remember that the two numbers appearing on the right of represent a power of 10 in scientific notation.

31. (a) Are the selections ordered or unordered? That is the question!

(b) That is always (well often at least) the question!

33. (a) The two members of a duet have equal standing so the order that they are chosen to be a part of the duet doesn't matter.

(b) Suppose that "The Revolution" were one person. Would Prince backed by "The Revolution" sound the same as "The Revolution" backed by Prince?

D. **General Probability Spaces**

35. The probabilities of the outcomes must sum to 1 in order to have a probability space.

39. Not all of the outcomes are equally likely. The probability of landing in a particular sector is proportional to the size of that sector.

E. **Events**

41. (a) $E_1 = \{HHT, \ldots\}$

(b) E_2 consists of two outcomes. List them.

(c) E_3 doesn't have much in it. But it can still be written using set notation. Remember that an event is a *set* of outcomes.

45. (b) The last toss could be tails or the second to last toss could be tails or the third to last toss could be tails or…

(c) 7 heads and 3 tails? 6 heads and 4 tails? Hmm…

F. Equiprobable Spaces

47. Exercise 41 should help you with the numerators. The size of the entire (equiprobable) sample space will help you with the denominator.

51. Again, Exercise 45 should help you with the numerators. The size of the entire (equiprobable) sample space will help you with the denominator.

55. You have 2 choices on the first problem (T or F), 2 choices on the second problem, etc. The total number of outcomes in this random experiment (the denominator in these problems) is found by the multiplication rule.

    (a) There is only one way to get all ten correct. The key must match the answer sheet.

    (b) There is only one way to get all ten incorrect (and hence a score of -5). The key must completely disagree with the answer sheet.

    (c) In order to get 8.5 points, the student must get exactly 9 correct answers and 1 incorrect answer. Any one of 10 different questions could be answered incorrectly.

    (d) In order to get 8 or more points, the student must get at least 9 correct answers (if they only get 8 correct answers, they lose a point for guessing two incorrect answers and score 7 points).

    (e) If the student gets 7 answers correct, they score $7 - 3 \times 0.5 = 5.5$ points. How can they score exactly 5 points?

    (f) If the student gets 8 answers correct, they score $8 - 2 \times 0.5 = 7$ points. How

many ways are there to select which 8 answers would be answered correctly (or, for that matter, which 2 would be answered incorrectly)?

57. (a) How many ways are there to choose four delegates? Assuming Alice is chosen, how many ways are there to choose the other three delegates?

    (b) Part (a) will come in handy here.

G. Odds

59. (a) If the probability of event E is a/b than the odds in favor of E are a to $b-a$.

    (b) Since 0.6 is rational, it can be written as a fraction.

61. If the odds in favor of event E are m to n than the probability of E is $m/(m+n)$.

JOGGING

63. Select the chair first and then the secretary. At the next stage, select the other three members.

65. (a) The outcomes in this event must be "of length 5" and "end in X winning."

    (b) The outcomes in this event must be "of length 5."

67. (b) A circle of 10 people can be broken to form a line in 10 different ways.

    (c) There are 2 choices as to which sex will start the line. Then, count how many ways there are to order the boys (the next stage) and how many ways there are to order the girls in the line (the last stage).

69. Form the groups in three stages.

71. (a) The answer is *not* ½.

(b) It will be useful to count how many ways one can choose the positions of the H's.

(c) $\Pr(3 \text{ or more } H\text{'s}) = 1 - \Pr(0 \ H\text{'s}) - \Pr(1 \ H\text{'s}) - \Pr(2 \ H\text{'s})$

73. The total number of 5-card *draw* poker hands (see Example 15.12) is 2,598,960. How many hands have all 5 cards the same color? [The first card can be any color, the remaining cards must match that color.]

75. Count the number of ways to get 10, J, Q, K, A of any suit (including all the same suit) and subtract the number of ways for these cards to all be the same suit.

77. $\Pr(\text{win}) = 1 - \Pr(\text{never roll a } 7)$

79. **(a)** This is the event that the first stereo is not defective and the second is not defective and the third is not defective and...

(b) $\Pr(\text{at most 1 defective}) = \Pr(\text{none defective}) + \Pr(1 \text{ defective})$

Chapter 16

WALKING

A.   Normal Curves

1. **(a)** Remember that in a normal distribution, the mean and the median is located at the center of the distribution.

(c) In a normal distribution, the standard deviation equals the distance between a point of inflection and the (vertical) axis of symmetry for the distribution.

3. **(a)** $Q_1 = \mu - 0.675\sigma$

(b) $Q_3 = \mu + 0.675\sigma$

5. First, find the mean μ. Then use the hint for Exercise 3(b).

7. Use your algebra skills to solve the equation $Q_3 = \mu + 0.675\sigma$ for μ.

11. See the hint for Exercise 1(a).

13. Normal distributions are very symmetric.

B.   Standardizing Data

15. **(a)** For a normal distribution with mean μ and standard deviation σ, a data value x has a standardized value z obtained by subtracting the mean μ from x and dividing the result by the standard deviation σ. In other words,
$$z = \frac{x - \mu}{\sigma}.$$

(b) Since the units of $x - \mu$ (kg) and the units of σ (kg) are the same, the value of z has no units.

17. **(a)** Use $Q_3 \approx \mu + (0.675)\sigma$ to first determine the value of σ.

21. Solving $z = \frac{x - \mu}{\sigma}$ for x gives $x = \mu + \sigma \cdot z$.

23. Solve $z = \frac{x - \mu}{\sigma}$ for σ.

25. Solve $z = \frac{x - \mu}{\sigma}$ for μ.

27. Solve for μ and for σ:
$$\frac{20 - \mu}{\sigma} = -2; \frac{100 - \mu}{\sigma} = 3$$

C.   The $68 - 95 - 99.7$ Rule

29. In every normal distribution, approximately 68% of all the data values fall within 1 standard deviation above and below the mean. Also, 95% of all the data values fall within 2 standard deviations above and below the mean. Approximately 99.7% of all the data values fall within 3 standard deviations above and below the mean. Moreover, 50% of the data lies between the quartiles.

31. See the hint for Exercise 29.

35. **(a)** Since 95% of the data lies within two standard deviations of the mean, 2.5% of the data are not within two standard deviations on each side of the mean.

    (b) This is half of the amount of data that falls between 1 and 2 standard deviations of the mean.

37. In a normal distribution, approximately 84% of the data lies above one standard deviation of the mean.

39. Convert these values to standardized values in order to determine their percentiles.

D. **Approximately Normal Data Sets**

41. **(c)** 41 has a standardized value of -1 and 63 has a standardized value of 1.

    (d) 63 is one standard deviation above the mean. Also, 68% of the data is within one standard deviation of the mean (so that 34% either more than one standard deviation above the mean or less than one standard deviation below the mean).

45. **(a)** 99 is two standard deviations below the mean; 151 is two standard deviations above the mean. Use the 68-95-99.7 rule.

    (b) 112 is one standard deviation below the mean. 151 is two standard deviations

above the mean. The answer, then, is certainly more than 68% of the patients but less than 95% of the patients.

47. A value 2 standard deviations below the mean is at about the 3rd percentile. A value 1 standard deviation below the mean is at about the 16th percentile. A value 1 standard deviation above the mean is at about the 84th percentile and a value 2 standard deviations above the mean is at about the 97th percentile.

53. **(a)** 15.25 has a standardized value of -1.

    (b) 21.25 has a standardized value of 2.

    (c) the 75th percentile is .0675 standard deviations above the mean.

E. **The Honest and Dishonest Coin Principles**

57. **(a)** According to the honest-coin principle, the random variable Y is approximately normal with mean $\mu = \dfrac{n}{2}$ and standard deviation $\sigma = \dfrac{\sqrt{n}}{2}$.

    (b) 1770 is approximately 1 standard deviation below the mean.

59. According to the honest-coin principle, the number of females in the sample is approximately normal with mean $\mu = \dfrac{n}{2}$ and standard deviation $\sigma = \dfrac{\sqrt{n}}{2}$. Convert all the given numerical values to standardized values.

61. If p is the probability of the coin landing heads, then X has an approximately normal distribution with mean $\mu = np$ and standard deviation $\sigma = \sqrt{np(1-p)}$.

63. The die is behaving like a dishonest coin with $p = 1/6$. There are two possible outcomes (a "6" and not a "6").

65. We assume that the defects are distributed normally so that the mean number of defects on a given day is 9,000 (since the mean satisfies $\mu = np$). Use the dishonest coin principle to find the standard deviation $\sigma = \sqrt{np(1-p)}$ and determine how many standard deviations 9,180 is above the mean.

JOGGING

67. (a) According to the table, the 95th percentile has a value of $\mu + 1.65\sigma$.

(b) According to the table, the 40th percentile has a value of $\mu - 0.25\sigma$.

69. Work in terms of standardized values.

73. (a) According to the table, the 90th percentile of the data is located at $\mu + 1.28 \times \sigma$ points.

(b) The 70th percentile of the data is located at $\mu + 0.52 \times \sigma$ points.

75. There is a 95% chance when X is within 2 standard deviations of the mean $\mu = \dfrac{n}{2}$. So,

$$15 = 2\left(\frac{\sqrt{n}}{2}\right).$$

77. Place $p = \dfrac{1}{2}$ in the dishonest-coin principle.

Chapter 1

Learning Outcomes

A successful student can

➡ construct and interpret a preference schedule for an election involving preference ballots.

➡ implement the plurality, Borda count, plurality-with-elimination, and pairwise comparisons vote counting methods.

➡ rank candidates using recursive and extended methods.

➡ identify fairness criteria as they pertain to voting methods.

➡ understand the significance of Arrows' impossibility theorem.

Skills to Help Prepare You for the Next Exam

At a <u>minimum,</u> be able to

➡ construct and interpret preference schedules for elections involving preference ballots (Exercises 1-10).

➡ determine the winner of a election using preference ballots using the
 - plurality method (Exercises 11, 12).
 - Borda count method (Exercises 17-22).
 - plurality-with-elimination method (Exercises 27-34).
 - method of pairwise comparisons (Exercises 35-38).

➡ rank candidates using
 - extended methods (Exercises 41-44).
 - recursive methods (Exercises 45-50).

➡ state the fairness criteria and identify when they are violated (Exercises 17-20, 27, 33-35, 59, 60, 62).

➡ count the number of pairwise comparisons in a given election (Exercises 51-56), the number of votes needed to win an election (Exercises 13-16), and Borda points given out (Exercises 23-26, 66-71).

➡ state Arrows' impossibility theorem in your own words.

Study Tip

When it comes to implementing a particular algorithm, practice makes perfect. Practice will also make you more efficient at solving routine problems on exam day when time is short and your mind has other things to worry about.

Chapter 2

Learning Outcomes

A successful student can

➡ represent a weighted voting system using a mathematical model.

➡ calculate the Banzhaf and Shapley-Shubik power distribution in a weighted voting system.

Skills to Help Prepare You for the Next Exam

At a <u>minimum,</u> be able to

➡ effectively use weighted voting terminology (Exercises 1, 2, 7, 8).

➡ construct a mathematical model of a weighted voting system (Exercises 3-6, 33, 34, 50, 51, 66).

➡ identify the presence of dictators, veto power, and dummies in a weighted voting system (Exercises 7-10, 45, 46, 52-56)

➡ compute Banzhaf and Shapley-Shubik power indices
- by listing all (sequential) coalitions and identifying critical and pivotal players (Exercises 11-16, 23-26).
- using techniques other than enumeration of all (sequential) coalitions (Exercises 17-20, 27-34, 47-49, 54, 69, 67).

➡ compute factorials
- using a calculator (Exercises 35, 36).
- by hand (Exercises 37-40).

➡ determine possible values of the quota q for a weighted voting system (Exercises 3-6, 45, 46, 54).

➡ count the number of coalitions having a particular property (Exercises 22, 41-44, 49, 59)

Study Tip

It is often the case that a problem can be solved using more than one technique. For example, in theory one can always compute the Banzhaf and Shapley-Shubik power indices for a given weighted voting system by listing all critical and pivotal players. However, when the system has many voters, this process is (very!) long and tedious. Be on the lookout for special structure (e.g. equal number of votes, dictators, dummies, etc.) that will allow for other techniques which reduce the number of calculations needed.

Chapter 3

Learning Outcomes

A successful student can

➡ state the fair-division problem and identify assumptions used in developing solution methods.

➡ recognize the differences between continuous and discrete fair-division problems.

➡ apply the divider-chooser, lone-divider, lone-chooser, and last-diminisher methods to continuous fair-division problems.

➡ apply the method of sealed bids and the method of markers to discrete fair-division problems.

Skills to Help Prepare You for the Next Exam

At a <u>minimum,</u> be able to

➡ quantify players' value systems (Exercises 1-4, 11-14, 17, 18, 29, 30, 31-40, 47-50, 57, 58, 65-68).

➡ identify fair shares to a given player (Exercises 5-10).

➡ apply the following methods to solve continuous fair-division problems:
  - divider-chooser (Exercises 11-18)
  - lone-divider (Exercises 19-30, 69)
  - lone-chooser (Exercises 31-40)
  - last-diminisher (Exercises 41-50, 71-73)

➡ apply the following methods to solve discrete fair-division problems:
  - sealed bids (Exercises 51-58).
  - method of markers (Exercises 59-68).

➡ add, subtract, multiply and divide fractions.

Study Tip

Carefully reading a mathematics textbook is an art that you should master. Read slowly--every word counts! Understand what you read before forging ahead. Mathematical terms and symbols have a very specific meaning—be sure that you understand the meaning of all of them. Very little information is repeated so you will likely want to read the text more than once (before and after class!). Working through the examples on paper as you read can keep you from nodding off and will add insight into what you read.

Chapter 4

Learning Outcomes

A successful student can

➡ state the basic apportionment problem.

➡ implement the methods of Hamilton, Jefferson, Adams, and Webster to solve apportionment problems.

➡ state the quota rule and determine when it is satisfied.

➡ identify paradoxes when they occur.

➡ understand the significance of Balinski and Young's impossibility theorem.

Skills to Help Prepare You for the Next Exam

At a <u>minimum,</u> be able to

➡ compute standard divisors and quotas for a given apportionment problem.

➡ apply Hamilton's method (Exercises 11-18).

➡ apply Jefferson's and Adams's methods (Exercises 23-30, 33-40).

➡ apply Webster's method (Exercises 43-50).

➡ identify a paradox when it presents itself (Exercises 19-22).

➡ identify when the quota rule is violated (Exercises 31, 32, 41, 42, 55, 56).

➡ state Balinski and Young's impossibility theorem in your own words.

Study Tip

Taking time to write neatly and organize your thoughts can make all the difference in the world. This is especially true when the computations in a particular problem are straightforward. It often pays to slow down and construct a table in order to avoid careless errors.

Chapter 5

Learning Outcomes

A successful student can

- ➡ identify and model Euler circuit and Euler path problems.

- ➡ understand the meaning of basic graph terminology.

- ➡ classify which graphs have Euler circuits or paths using Euler's circuit theorems.

- ➡ implement Fleury's algorithm to find an Euler circuit or path when it exists.

- ➡ eulerize and semi-eulerize graphs when necessary.

- ➡ recognize an optimal eulerization (semi-eulerization) of a graph.

Skills to Help Prepare You for the Next Exam

At a <u>minimum</u>, be able to

- ➡ demonstrate an understanding of basic graph terminology (Exercises 1, 2, 7-14, 45, 53).

- ➡ draw a graph modeling a particular situation (Exercises 15-20).

- ➡ determine if a given graph has an Euler circuit or an Euler path (Exercises 21-26).

- ➡ find an Euler circuit or path for a graph having one (Exercises 27-36).

- ➡ find an (optimal) eulerization for a graph that has no Euler circuit (Exercises 41, 42, 51, 52, 62).

- ➡ find an (optimal) semi-eulerization for a graph that has no Euler path (Exercises 43, 44, 47, 48, 58, 59, 63).

Study Tip

A picture is worth a thousand words. A well-labeled diagram can lead to understanding many problems and can help solve many others. Try drawing diagrams even for problems that don't seem to lend themselves to this approach.

Chapter 6

Learning Outcomes

A successful student can

➡ identify and model Hamilton circuit and Hamilton path problems.

➡ recognize complete graphs and state the number of Hamilton circuits that they have.

➡ identify traveling-salesman problems and the difficulties faced in solving them.

➡ implement brute-force, nearest-neighbor, repeated nearest-neighbor, and cheapest-link algorithms to find approximate solutions to traveling-salesman problems.

➡ recognize the difference between efficient and inefficient algorithms.

➡ recognize the difference between optimal and approximate algorithms.

Skills to Help Prepare You for the Next Exam

At a <u>minimum,</u> be able to

➡ find one or more Hamilton circuits or Hamilton paths for a given graph (Exercises 1-8, 57-63).

➡ calculate factorials by hand and with a calculator (Exercises 17-24).

➡ calculate how many edges and Hamilton paths there are in a given complete graph (Exercises 25-28).

➡ use the brute-force algorithm to find optimal solutions to traveling-salesman problems stated in various contexts (Exercises 29-30).

➡ use the nearest-neighbor algorithm and repetitive nearest-neighbor algorithm to find approximate solutions to traveling-salesman problems stated in various contexts (Exercises 29-36, 37-42, 69).

➡ use the cheapest-link algorithm to find approximate solutions to traveling-salesman problems stated in various contexts (Exercises 43-48, 70).

➡ draw a graph model for a traveling-salesman problem (Exercises 51-56)

Study Tip

While calculators can handle many routine calculations, the ability to calculate by hand should not be dismissed. A great deal of understanding of mathematical concepts comes from cultivating this ability.

Chapter 7

Learning Outcomes

A successful student can

➡ identify and use a graph to model minimum network problems.

➡ classify which graphs are trees.

➡ implement Kruskal's algorithm to find a minimal spanning tree.

➡ understand Torricelli's construction for finding a Steiner point.

➡ recognize when the shortest network connecting three points uses a Steiner point.

➡ understand basic properties of the shortest network connecting a set of (more than three) points.

Skills to Help Prepare You for the Next Exam

At a <u>minimum</u>, be able to

➡ determine whether a graph is a tree or not (Exercises 1-10).

➡ find one or more or even all of the spanning trees of a given network (Exercises 11-18, 49).

➡ find a minimal spanning tree for a weighted network (Exercises 19-24).

➡ determine the shortest network connecting three points (Exercises 27-34).

➡ compute information about graphs containing 30-60- 90° triangles (Exercises 45-48, 63, 64).

➡ reproduce Torricelli's construction (with a straightedge and compass).

Study Tip

Whenever an algorithm (or fact, construction, theorem, etc.) is named after a person, it is almost always important enough to master. Practice it enough to feel comfortable applying it in an exam environment in which you don't have the aid of a textbook or notes.

Chapter 8

Learning Outcomes

A successful student can

➡ understand and use digraph terminology.

➡ schedule a project on N processors using the priority-list model.

➡ apply the backflow algorithm to find the critical path of a project.

➡ implement the decreasing-time and critical-path algorithms.

➡ recognize optimal schedules and the difficulties faced in finding them.

Skills to Help Prepare You for the Next Exam

At a <u>minimum,</u> be able to

➡ find indegrees, outdegrees, vertex-sets, arc-sets, cycles, etc. for a given digraph (Exercises 1-12, 51).

➡ use digraphs to model real-life situations (Exercises 13-22, 60).

➡ use a given priority list to schedule a project on N processors (Exercises 27-30, 35, 36).

➡ use the decreasing-time algorithm to schedule a project on N processors (Exercises 37-44).

➡ apply the backflow and critical-path algorithms (Exercises 45-50).

➡ schedule independent tasks and determine how close the result is to optimal (Exercises 51-58).

Study Tip

Never underestimate the power of common sense. It is always wise to check your answer to a problem to see if it makes sense. It is doesn't, sometimes it is best to simply start over from the very beginning.

Chapter 9

Learning Outcomes

A successful student can

- ➡ generate the Fibonacci sequence and identify some of its properties.

- ➡ identify relationships between the Fibonacci sequence and the golden ratio.

- ➡ define a gnomon and understand the concept of similarity.

- ➡ recognize gnomonic growth in nature.

Skills to Help Prepare You for the Next Exam

At a <u>minimum,</u> be able to

- ➡ use and understand subscript notation (Exercises 1-6, 13, 14, 27, 28, 51).

- ➡ state and apply the defining characteristics of the Fibonacci numbers (Exercises 7,8, 25, 26, 52-54).

- ➡ state and apply basic properties of the golden ratio $\left(e.g.\ \phi^2 = \phi + 1\right)$ (Exercises 23, 24, 55, 57, 58, 68).

- ➡ evaluate expressions in which order of operations is essential (Exercises 19-22).

- ➡ work with scientific notation (Exercises 25, 26).

- ➡ solve quadratic equations (Exercises 29-32).

- ➡ determine when two figures are similar (Exercises 35, 36, 39, 40).

- ➡ determine parameters of gnomons to given plane figures (Exercises 41-49)

- ➡ state the *exact* value of the golden ratio $\left(\phi = \dfrac{1+\sqrt{5}}{2}\right)$ and an approximate value (1.618).

Study Tip

There is great beauty in mathematics. It is very common that patterns emerge in solving mathematical problems. Keep an eye open for these patterns and take advantage of them in solving problems when the opportunity arises.

Chapter 10

Learning Outcomes

A successful student can

➡ understand how a transition rule models population growth.

➡ recognize linear, exponential, and logistic growth models.

➡ apply linear, exponential, and logistic growth models to solve population growth problems.

➡ differentiate between recursive and explicit models of population growth.

➡ apply the general compounding formula to answer financial questions.

➡ state and apply the arithmetic and geometric sum formulas in their appropriate contexts.

Skills to Help Prepare You for the Next Exam

At a <u>minimum,</u> be able to

➡ find a particular term of an arithmetic, geometric, or logistic sequence (Exercises 1-9, 17-26, 45-54).

➡ give a recursive and/or explicit description of a population sequence (Exercises 1-2, 5-6, 9-10, 17-26).

➡ identify a population model and/or find common differences/ratios (Exercises 6, 13-14, 20, 55, 58, 67, 72, 74).

➡ compute the sum of a given number of terms in an arithmetic or geometric sequence (Exercises 11-18, 25-26, 41-42, 64-66, 68-70).

➡ apply the general compounding formula (Exercises 29-36, 39, 41-44).

➡ solve problems involving percentage increases/decreases (Exercises 27-28, 62-63).

➡ compute annual yield (Exercises 35-40, 59, 60).

Study Tip

Before memorizing any formulas, it is always important to understand why they work. This knowledge will make remembering the formula when it is needed much easier.

Chapter 11

Learning Outcomes

A successful student can

➡ describe the basic rigid motions of the plane and state their properties.

➡ classify the possible symmetries of any finite two-dimensional shape or object.

➡ classify the possible symmetries of a border pattern.

Skills to Help Prepare You for the Next Exam

At a <u>minimum</u>, be able to

➡ find the image of a shape under a reflection given the
  - axis of reflection (Exercises 1-4).
  - location of the image of one point (Exercises 4-8).
  - location of two fixed points of the reflection (Exercises 9, 10).

➡ find the image of a shape under a rotation given the
  - location of the rotocenter and the angle of rotation (Exercises 11, 12).
  - location of the image of two points (Exercises 15-18).
  - image of one point and the angle of rotation (Exercises 19, 20).

➡ find the image of a shape under a translation given the
  - translation vector (Exercises 21, 22).
  - image of a point under the translation (Exercises 23, 24).

➡ find the image of a shape under a glide reflection given
  - a description of the translation and reflection (Exercises 27, 28).
  - the location of the image of two points (Exercises 29-34).

➡ give the symmetry type (D_N or Z_N) of finite figures in the plane (Exercises 39-44).

➡ give the symmetry type (mm, mg, m1, 1m, 1g, 12, 11) of border patterns (Exercises 49-52).

➡ identify a rigid motion based on its properties (Exercises 55, 56, 59-66, 68, 69, 71).

Study Tip

Don't overlook the role mnemonic devices play in memorization. For example, in classifying border patterns, the second symbol is selected from the following (ordered) list: mg21. Your friends might give you "**m**ore **g**rief when you turn **21**." There are even folks that drink "**M**iller **G**enuine" but, of course, not until they are **21**.

Chapter 12

Learning Outcomes

A successful student can

➡ explain the process by which fractals such as the Koch snowflake and the Sierpinski Gasket are constructed.

➡ recognize self-similarity (or symmetry of scale) and its relevance.

➡ describe how random processes can create fractals such as the Sierpinski Gasket.

➡ explain the process by which the Mandelbrot set is constructed.

Skills to Help Prepare You for the Next Exam

At a <u>minimum,</u> be able to

➡ calculate the perimeter and area of the figure found at any step in the construction of a given fractal (Exercises 1-14, 16-20, 23, 24, 26, 27, 29, 30, 58).

➡ draw the outcome of a particular step in the construction of a fractal (Exercises 15, 25, 28, 57).

➡ play the chaos game (Exercises 33-40).

➡ add and multiply complex numbers (Exercises 41-44).

➡ construct the Mandelbrot sequence for a given seed *s* and identify it as escaping, periodic, or attracted (Exercises 47-52).

Study Tip

Spotting patterns is essential in mathematics. It takes patience. It is also often useful to *not* simply algebraic expressions and to resist the temptation to reach for a calculating device in the search for a pattern.

Chapter 13

Learning Outcomes

A successful student can

- ➡ identify whether a given survey or poll is biased.

- ➡ list and discuss the quality of several sampling methods.

- ➡ identify components of a well-constructed clinical study.

- ➡ define key terminology in the data collection process.

- ➡ estimate the size of a population using the capture-recapture method.

Skills to Help Prepare You for the Next Exam

At a <u>minimum</u>, be able to

- ➡ identify the population, sample, and sampling frame (Exercises 1, 4, 6, 9, 11, 13, 17, 21, 39, 43, 48, 55).

- ➡ distinguish between a parameter and a statistic (Exercises 2, 3, 5, 13, 26, 59).

- ➡ identify the sampling method used in a survey or poll (Exercises 1, 6, 10, 13, 19, 23, 25, 27, 28).

- ➡ identify whether sampling error is a result of sampling variability or bias (Exercises 3, 8, 11, 19, 23).

- ➡ estimate the size of a population using capture-recapture (Exercises 29-36).

- ➡ identify whether a study is blind, double-blind, or neither (Exercises 45, 49).

- ➡ list possible confounding variables in an experiment (Exercises 40, 49, 55-57).

- ➡ discuss elements of poor design (Exercises 15, 20, 24, 41, 50, 55, 56, 61, 63, 64).

Study Tip

Mathematics is not just a bunch of equations and formulas. Your instructor will expect you to communicate ideas in a manner similar to what you may have encountered in high school English class. When writing a paper, even in math class, always use complete sentences and proper grammar just as you would in any other class.

Chapter 14

Learning Outcomes

A successful student can

➡ interpret and produce an effective graphical summary of a data set.

➡ identify various types of numerical variables.

➡ interpret and produce numerical summaries of data including percentiles and five-number summaries.

➡ describe the spread of a data set using range, interquartile range, and standard deviation.

Skills to Help Prepare You for the Next Exam

At a <u>minimum</u>, be able to

➡ read and produce frequency tables, bar graphs, and pie charts (Exercises 1-16).

➡ read and interpret histograms (Exercises 19-22).

➡ compute means and medians (Exercises 23-32, 65, 66).

➡ compute percentiles (including quartiles) (Exercises 33-40).

➡ compute five-number summaries and produce box plots (Exercises 41-44).

➡ interpret box plots (Exercises 45, 46).

➡ compute range, interquartile range, and determine the existence of outliers (Exercises 47-54).

➡ compute standard deviation (Exercises 55-58, 72, 80).

➡ compute mode (Exercise set 59-64).

Study Tip

When you feel as if you have mastered a particular computational skill, put yourself in an exam environment and see how you fare. If all goes well, focus your attention on problems that appear to be stated a bit differently or are less computational (theoretical) in nature.

Chapter 15

Learning Outcomes

A successful student can

- ➡ describe an appropriate sample space of a random experiment.

- ➡ apply the multiplication rule, permutations, and combinations to counting problems.

- ➡ understand the concept of a probability assignment.

- ➡ identify independent events and their properties.

- ➡ use the language of odds in describing probabilities of events.

Skills to Help Prepare You for the Next Exam

At a <u>minimum</u>, be able to

- ➡ describe an appropriate sample space for a given random experiment (Exercises 1-8).

- ➡ apply the multiplication rule to counting problems (Exercises 9-18, 66, 67).

- ➡ compute the value of a given combination or permutation (Exercises 19-23, 29, 30).

- ➡ apply combinations or permutations to counting problems (Exercises 31-34, 63).

- ➡ construct a probability assignment for a given sample space (Exercises 35-40).

- ➡ describe an event using set notation (Exercises 41-46, 64, 65).

- ➡ find the probability of a given event (Exercises 47-58, 70-80).

- ➡ write odds as probabilities and vice-versa (Exercises 59-62).

Study Tip

Probability is a topic that can produce answers that are not intuitive. It can also consist of problems that are very easy to state but difficult to solve. In calculating probabilities be sure to write out an appropriate sample space for the random experiment. Don't attempt to solve the entire problem in your head.

Chapter 16

Learning Outcomes

A successful student can

➡ identify and describe an approximately normal distribution.

➡ state properties of a normal distribution.

➡ understand a data set in terms of standardized data values.

➡ state the 68-95-99.7 rule.

➡ apply the honest and dishonest-coin principles to understand the concept of a confidence interval.

Skills to Help Prepare You for the Next Exam

At a <u>minimum,</u> be able to

➡ find parameters (mean, median, standard deviation, etc.) of a normal distribution given in graphical form (Exercises 1, 2, 9, 10).

➡ find the value of other parameters given two parameters of a normal distribution (Exercises 3-8).

➡ find standardized values given two parameters of a normal distribution (Exercises 15-18).

➡ use standardized values to determine parameters of a normal distribution (Exercises 21-28).

➡ determine the percent of data that falls between key values of a normal distribution (Exercises 29-40).

➡ apply the 68-95-99.7 rule to approximately normal distributions (Exercises 41-52).

➡ use the (dis)honest-coin principle to understand a random variable X (Exercises 57-66, 71).

Study Tip

A well-prepared student will not be surprised by exam questions. Solving as many different types of problems as possible can help you prepare much more than memorizing a particular example or equation. Mathematics is not a spectator sport. The best way to learn mathematics is to do mathematics.

Chapter 1

WALKING

A. **Ballots and Preference Schedules**

1. (a) There are 12 votes all together. A majority is more than half of the votes, or at least 7.

(b) The Country Cookery has the most first-place votes (6). Since this is not more than half of the first-place votes, it is a plurality.

(c)

Number of voters	5	3	1	3
1st choice	A	C	B	C
2nd choice	B	B	D	B
3rd choice	C	A	C	D
4th choice	D	D	A	A

3. (a) $5 + 3 + 5 + 3 + 2 + 3 = 21$

(b) There are 21 votes all together. A majority is more than half of the votes, or at least 11.

(c) A $(5 + 3 = 8)$

(d) E $(5 + 3 + 2 = 10)$

5. (a) No candidate has a majority of the first-place votes. So all candidates with 20% or less of the 21 first-place votes are eliminated. $0.20(21) = 4.2$, so all candidates with fewer than 4.2 first-place votes (4 votes or fewer) are eliminated. The candidates that are eliminated are B (3 first-place votes) and E (0 first-place votes).

(b)

Number of voters	5	3	5	3	2	3
1st choice	A	A	C	D	D	A
2nd choice	C	D	D	C	C	C
3rd choice	D	C	A	A	A	D

This schedule can be more efficiently written as follows:

Number of voters	8	3	5	5
1st choice	A	A	C	D
2nd choice	C	D	D	C
3rd choice	D	C	A	A

(c) Candidate A now has 11 first-place votes and is the majority winner.

7.

Number of voters	255	480	765
1st choice	L	C	M
2nd choice	M	M	L
3rd choice	C	L	C

$(0.17)(1500) = 255$; $(0.32)(500) = 480$; The remaining voters (51% or 1500-255-480=765) prefer M the most, C the least, so that L is their second choice.

9.

Number of voters	47	36	24	13	5
1st choice	B	A	B	E	C
2nd choice	E	B	A	B	E
3rd choice	A	D	D	C	A
4th choice	C	C	E	A	D
5th choice	D	E	C	D	B

B.   Plurality Method

11. (a) A has $153 + 102 + 55 = 310$ first-place votes.
B has $202 + 108 + 20 = 330$ first-place votes.
C has $110 + 160 = 270$ first-place votes.
D has $175 + 155 = 330$ first-place votes.
B and D tie with 330 first-place votes each.

(b) B has $55 + 110 + 175 = 340$ last-place votes.
D has $153 + 20 + 160 = 333$ last-place votes.
D wins the election.

13. (a) 23 votes will guarantee A at least a tie for first; 24 votes guarantee that A is the only winner. (With 23 of the remaining 30 votes A has 49 votes. The only candidate with a chance to have that many votes is C. Even if C gets the other 7 remaining votes, C would not have enough votes to beat A.)

For another way to compute this answer, suppose that A receives x out of the remaining 30 votes and the competitor with the most votes, C, receives the remaining 30 - x votes. To determine the values of x that guarantee a win for A, we solve $26 + x \geq 42 + (30 - x)$, i.e. $2x \geq 46$. So $x \geq 23$ votes will guarantee A at least a tie for first.

(b) 11 votes will guarantee C at least a tie for first; 12 votes guarantee C is the only winner. (With 11 of the remaining votes C has 53 votes. The only candidate with a chance to have that many votes is D. Even if D gets the other 19 remaining votes, D would not have enough votes to beat C.)

For another way to compute this answer, suppose that C receives y out of the remaining 30 votes and the competitor with the most votes, D, receives the remaining 30 - y votes. To determine the values of y that guarantee a win for C, we solve $42 + y \geq 34 + (30 - y)$, i.e. $2y \geq 22$. So $y \geq 11$ votes will guarantee C at least a tie for first.

15. (a) $721/2 = 360.5$ so that 361 votes are needed to have a majority.

(b) The winning candidate has the smallest number of votes possible if the votes are distributed as evenly as possible. $\frac{721}{5} = 144.2,$ so the smallest number of votes a winning candidate can have is 145. (If this were the case, then four of the candidates each receive 144 votes and one of the candidates, the winning candidate, receives 145 votes.)

(c) The winning candidate has the smallest number of votes possible if the votes are distributed as evenly as possible. $\frac{721}{10} = 72.1,$ so the smallest number of votes a winning candidate can have is 73.

(If this were the case, then nine of the candidates each receive 72 votes and one of the candidates, the winning candidate, receives 73 votes.)

C.   Borda Count Method

17. (a) *A* has $5 \times (5+3) + 4 \times 0 + 3 \times 3 + 2 \times (5+2) + 1 \times 3 = 66$ points.

B has $5 \times 3 + 4 \times 5 + 3 \times (3+3+2) + 2 \times 0 + 1 \times 5 = 64$ points.

C has $5 \times 5 + 4 \times (3+2) + 3 \times 5 + 2 \times (3+3) + 1 \times 0 = 72$ points.

D has $5 \times (3+2) + 4 \times 3 + 3 \times 5 + 2 \times 5 + 1 \times 3 = 65$ points.

E has $5 \times 0 + 4 \times (5+3) + 3 \times 0 + 2 \times 3 + 1 \times (5+3+2) = 48$ points.

The winner is Professor Chavez.

(b)

Number of voters	5	3	5	5	3
1st choice	*A*	*A*	*C*	*D*	*B*
2nd choice	*B*	*D*	*D*	*C*	*A*
3rd choice	*C*	*B*	*A*	*B*	*C*
4th choice	*D*	*C*	*B*	*A*	*D*

A has $4 \times (5+3) + 3 \times 3 + 2 \times 5 + 1 \times 5 = 56$ points.

B has $4 \times 3 + 3 \times 5 + 2 \times (3+5) + 1 \times 5 = 48$ points.

C has $4 \times 5 + 3 \times 5 + 2 \times (5+3) + 1 \times 3 = 54$ points.

D has $4 \times 5 + 3 \times (3+5) + 2 \times 0 + 1 \times (5+3) = 52$ points.

The winner is Professor Argand.

(c) Professor Chavez (choice *C*) was the winner of an election and when an irrelevant alternative (candidate *E*) was disqualified and the ballots recounted, we found that Professor Argand (choice *A*) was the winner of the reelection.

19. (a) *A* has $5 \times 8 + 4 \times (2+1) + 3 \times 7 + 2 \times 0 + 1 \times 6 = 79$ points.

B has $5 \times 0 + 4 \times (8+7+6) + 3 \times 2 + 2 \times 1 + 1 \times 0 = 92$ points.

C has $5 \times 2 + 4 \times 0 + 3 \times 8 + 2 \times (7+6) + 1 \times 1 = 61$ points.

D has $5 \times (7+6) + 4 \times 0 + 3 \times 1 + 2 \times (8+2) + 1 \times 0 = 88$ points.

E has $5 \times 1 + 4 \times 0 + 3 \times 6 + 2 \times 0 + 1 \times (8+7+2) = 40$ points.

The winner is Borrelli.

    (b) Dante has a majority of the first-place votes (13 out of a possible 24 votes) but does not win the election.

    (c) Dante, having a majority of the first-place votes, is a Condorcet candidate but does not win the election.

21. (a) Column 1: $0.40 \times 200 = 80$ voters; Column 2: $0.25 \times 200 = 50$ voters; Column 3: $0.20 \times 200 = 40$ voters; Column 4: $0.15 \times 200 = 30$ voters.

    A has $4 \times 80 + 3 \times 30 + 2 \times 40 + 1 \times 50 = 540$ points.

    B has $4 \times (40 + 30) + 3 \times 50 + 2 \times 80 + 1 \times 0 = 590$ points.

    C has $4 \times 50 + 3 \times 0 + 2 \times 0 + 1 \times (80 + 40 + 30) = 350$ points.

    D has $4 \times 0 + 3 \times (80 + 40) + 2 \times (50 + 30) + 1 \times 0 = 520$ points.

    The winner is B.

    (b) Column 1: $0.40 \times 20N = 8N$ voters; Column 2: $0.25 \times 20N = 5N$ voters; Column 3: $0.20 \times 20N = 4N$ voters; Column 4: $0.15 \times 20N = 3N$ voters.

    A has $4 \times 8N + 3 \times 3N + 2 \times 4N + 1 \times 5N = 54N$ points.

    B has $4 \times (4N + 3N) + 3 \times 5N + 2 \times 8N + 1 \times 0N = 59N$ points.

    C has $4 \times 5N + 3 \times 0N + 2 \times 0N + 1 \times (8N + 4N + 3N) = 35N$ points.

    D has $4 \times 0N + 3 \times (8N + 4N) + 2 \times (5N + 3N) + 1 \times 0N = 52N$ points.

    The winner is B.

    (c) No. The number of points for each candidate is just multiplied by N.

23. (a) To achieve the maximum number of points possible, a candidate would need to receive all 50 first-place votes. Such a candidate would have $4 \times 50 + 3 \times 0 + 2 \times 0 + 1 \times 0 = 200$ points.

    (b) To achieve the minimum number of points possible, a candidate would need to receive all 50 last-place votes. Such a candidate would have $4 \times 0 + 3 \times 0 + 2 \times 0 + 1 \times 50 = 50$ points.

25. (a) $4 + 3 + 2 + 1 = 10$ points

    (b) $\dfrac{10 \text{ points}}{1 \text{ ballot}} \times 110 \text{ ballots} = 1100 \text{ points}$

    (c) $1100 - 320 - 290 - 180 = 310$ points

D.   **Plurality-with-Elimination Method**

27. (a) Professor Argand

    Round 1:

Candidate	A	B	C	D	E
Number of first-place votes	8	3	5	5	0

    E is eliminated.

    Round 2: Eliminating E does not affect the first-place votes of the candidates. Using the table from Round 1, we see that B now has the fewest number of first-place votes, so B is eliminated.

Round 3: The three votes originally going to *B* would next go to *E*, but *E* has also been eliminated. So *B*'s three votes go to *A*.

Candidate	*A*	*B*	*C*	*D*	*E*
Number of first-place votes	11		5	5	

Candidate *A* now has a majority of the first-place votes and is declared the winner.

(b) Professor Epstein.

Since Professor Chavez is not under consideration, the resulting preference schedule would contain only four candidates.

Number of voters	5	3	5	3	2	3
1st choice	*A*	*A*	*E*	*D*	*D*	*B*
2nd choice	*B*	*D*	*D*	*B*	*B*	*E*
3rd choice	*D*	*B*	*A*	*E*	*A*	*A*
4th choice	*E*	*E*	*B*	*A*	*E*	*D*

Round 1:

Candidate	*A*	*B*	*D*	*E*
Number of first-place votes	8	3	5	5

B is eliminated.

Round 2: The three votes originally going to *B* go to *E*.

Candidate	*A*	*B*	*D*	*E*
Number of first-place votes	8		5	8

D is eliminated.

Round 3: The 5 votes that *D* had in round 2 would go to *B* except that *B* has been eliminated. Instead, three of these votes go to *E* and two of these five votes go to *A*.

Candidate	*A*	*B*	*D*	*E*
Number of first-place votes	10			11

Candidate *E* now has a majority of the first-place votes and is declared the winner.

(c) *A* was the winner of the original election, and, in the reelection without candidate *C* (an irrelevant alternative) it turned out that candidate *E* was declared the winner. This violates the Independence-of-Irrelevant-Alternatives criterion.

29. (a) Dante is the winner.
Round 1:

Candidate	*A*	*B*	*C*	*D*	*E*
Number of first-place	8	0	2	13	1

Candidate *D* has a majority of the first-place votes and is declared the winner.

(b) Since Dante is a majority winner, in this case the winner is determined in the first round.

(c) If there is a choice that has a majority of the first-place votes, then that candidate will be the winner under the plurality-with-elimination method. So, the plurality-with-elimination method satisfies the majority criterion.

31. B is the winner.

Round 1:

Candidate	A	B	C	D
Number of first-place	40	35	25	0

D is eliminated.

Round 2: Eliminating *D* does not affect the first-place votes of the other candidates. Using the table from Round 1, we see that candidate *C* now has the fewest number of first-place votes, so *C* is eliminated.

Round 3: The 25 votes originally going to *C* now go to *B*.

Candidate	A	B	C	D
Number of first-place	40	60		

Candidate *B* now has a majority of the first-place votes and is declared the winner.

33. **(a)** Año Nuevo, California is the winner.

Round 1:

Candidate	A	B	C	D
Number of first-place	10	11	4	2

Candidate *D* is eliminated.

Round 2: The two votes originally going to *D* now go to *C*.

Candidate	A	B	C	D
Number of first-place	10	11	6	

Candidate *C* is eliminated.

Round 3: Of the six votes going to *C* in Round 2, four will go to *A* and two will go to *B*.

Candidate	A	B	C	D
Number of first-place	14	13		

Candidate *A* has a majority of the first-place votes and is declared the winner.

(b) *C* is the Condorcet candidate.

In a head-to-head contest, *C* beats *A*, 17 votes to 10 votes.
In a head-to-head contest, *C* beats *B*, 16 votes to 11 votes.
In a head-to-head contest, *C* beats *D*, 19 votes to 8 votes.

(c) Cloudbreak, Fiji, which is the Condorcet candidate, fails to win the election under the plurality-with-elimination method. So, the plurality-with-elimination method fails to satisfy the Condorcet criterion.

E.   Pairwise Comparisons Method

35. **(a)** Chad is the winner.

A versus *B*: 13 votes to 13 votes (tie). *A* gets $\frac{1}{2}$ point, *B* gets $\frac{1}{2}$ point.

A versus *C*: 8 votes to 18 votes (*C* wins). *C* gets 1 point.
A versus *D*: 14 votes to 12 votes (*A* wins). *A* gets 1 point.
A versus *E*: 21 votes to 5 votes (*A* wins). *A* gets 1 point.
B versus *C*: 8 votes to 18 votes (*C* wins). *C* gets 1 point.
B versus *D*: 8 votes to 18 votes (*D* wins). *D* gets 1 point.
B versus *E*: 8 votes to 18 votes (*E* wins). *E* gets 1 point.

C versus *D*: 13 votes to 13 votes (tie). *C* gets $\frac{1}{2}$ point, *D* gets $\frac{1}{2}$ point.

C versus *E*: 13 votes to 13 votes (tie). *C* gets $\frac{1}{2}$ point, *E* gets $\frac{1}{2}$ point.

D versus *E*: 15 votes to 11 votes. *D* gets 1 point.

The final tally, is $2\frac{1}{2}$ points for *A*, $\frac{1}{2}$ point for *B*, 3 points for *C*, $2\frac{1}{2}$ points for *D*, and $1\frac{1}{2}$ points for *E*. Candidate *C*, Chad, is the winner.

(b) The preference schedule without Alberto is shown below.

Number of voters	8	6	5	5	2
1st choice	*C*	*E*	*E*	*D*	*D*
2nd choice	*B*	*D*	*C*	*C*	*E*
3rd choice	*D*	*B*	*D*	*E*	*B*
4th choice	*E*	*C*	*B*	*B*	*C*

B versus *C*: 8 votes to 18 votes (*C* wins). *C* gets 1 point.
B versus *D*: 8 votes to 18 votes (*D* wins). *D* gets 1 point.
B versus *E*: 8 votes to 18 votes (*E* wins). *E* gets 1 point.

C versus *D*: 13 votes to 13 votes (tie). *C* gets $\frac{1}{2}$ point, *D* gets $\frac{1}{2}$ point.

C versus *E*: 13 votes to 13 votes (tie). *C* gets $\frac{1}{2}$ point, *E* gets $\frac{1}{2}$ point.

D versus *E*: 15 votes to 11 votes. *D* gets 1 point.

The final tally, is 0 points for *B*, 2 points for *C*, $2\frac{1}{2}$ points for *D*, and $1\frac{1}{2}$ points for *E*. Candidate *D*, Dora, is now the winner.

(c) *C* was the winner of the original election, and, in the re-election without candidate *A* (an irrelevant alternative) it turned out that candidate *D* was declared the winner. This violates the Independence-of-Irrelevant-Alternatives criterion.

37. *A* versus *B*: 13 votes to 8 votes (*A* wins). *A* gets 1 point.
    A versus *C*: 11 votes to 10 votes (*A* wins). *A* gets 1 point.
    A versus *D*: 11 votes to 10 votes (*A* wins). *A* gets 1 point.
    A versus *E*: 10 votes to 11 votes (*E* wins). *E* gets 1 point.
    B versus *C*: 11 votes to 10 votes (*B* wins). *B* gets 1 point.
    B versus *D*: 8 votes to 13 votes (*D* wins). *D* gets 1 point.
    B versus *E*: 16 votes to 5 votes (*B* wins). *B* gets 1 point.
    C versus *D*: 13 votes to 8 votes (*C* wins). *C* gets 1 point.

C versus E: 18 votes to 3 votes (C wins). C gets 1 point.
D versus E: 13 votes to 8 votes (D wins). D gets 1 point.
The final tally is 3 points for A, 2 points for B, 2 points for C, 2 points for D, and 1 point for E.
Candidate A, Professor Argand, is the winner.

39. With five candidates, there are a total of 10 pairwise comparisons. E wins $10 - 2 - 2\frac{1}{2} - 1 - 1\frac{1}{2} = 3$. The 10 points are distributed as follows: A gets 2 points, B gets $2\frac{1}{2}$ points, C gets 1 point, D gets $1\frac{1}{2}$ points, and E gets the remaining 3 points. Candidate E is the winner.

F.   Ranking Methods

41. (a) A has $4 + 8 = 12$ first-place votes.
B has 1 first-place vote.
C has 9 first-place votes.
D has 5 first-place votes.
Winner A: Second place: C. Third place: D. Last place: B.

(b) A has $4 \times (4+8) + 3 \times 1 + 2 \times 9 + 1 \times 5 = 74$ points.
B has $4 \times 1 + 3 \times 8 + 2 \times (4+5) + 1 \times 9 = 55$ points.
C has $4 \times 9 + 3 \times (4+5) + 2 \times 0 + 1 \times (1+8) = 72$ points.
D has $4 \times 5 + 3 \times 9 + 2 \times (1+8) + 1 \times 4 = 69$ points.
Winner: A. Second place: C. Third place: D. Last place: B.

(c) Round 1:

Candidate	A	B	C	D
Number of votes	12	1	9	5

B is eliminated.
Round 2: The vote originally going to B now goes to A.

Candidate	A	B	C	D
Number of votes	13		9	5

D is eliminated.
Round 3: The five votes going to D now go to C.

Candidate	A	B	C	D
Number of votes	13		14	

A is eliminated, and C is declared the winner.
Winner: C. Second place: A. Third place: D. Last place: B.

(d) A versus B: 21 votes to 6 votes (A wins). A gets 1 point.
A versus C: 13 votes to 14 votes (C wins). C gets 1 point.
A versus D: 13 votes to 14 votes (D wins). D gets 1 point.
B versus C: 9 votes to 18 votes (C wins). C gets 1 point.
B versus D: 13 votes to 14 votes (D wins). D gets 1 point.
C versus D: 13 votes to 14 votes (D wins). D gets 1 point.

The final tally is 1 point for *A*, 0 points for *B*, 2 points for *C*, and 3 points for *D*.
Winner: *D*. Second place: *C*. Third place: *A*. Last place: *B*.

43.   **(a)**   *A* has 40% of the first-place votes.
         B has 35% of the first-place votes.
         C has 25% of the first-place votes.
         D has 0% of the first-place votes.
         Winner *A*: Second place: *B*. Third place: *C*. Last place: *D*.

   (b)   We treat each 1% of voters as 1 "block" of votes.

A has $4 \times 40 + 3 \times 15 + 2 \times 20 + 1 \times 25 = 270$ points.

B has $4 \times (20 + 15) + 3 \times 25 + 2 \times 40 + 1 \times 0 = 295$ points.

C has $4 \times 25 + 3 \times 0 + 2 \times 0 + 1 \times (40 + 20 + 15) = 175$ points.

D has $4 \times 0 + 3 \times (40 + 20) + 2 \times (25 + 15) + 1 \times 0 = 260$ points.

Winner: *B*. Second place: *A*. Third place: *D*. Last place: *C*.

   (c)   Round 1:

Candidate	*A*	*B*	*C*	*D*
Number of votes	40%	35%	25%	0%

D is eliminated.
Round 2: Eliminating *D* does not affect the first-place votes of the other candidates. Using the table from Round 1, we see that candidate *C* now has the fewest number of first-place votes, so *C* is eliminated.
Round 3: The 25% of the votes going to *C* now go to *B*.

Candidate	*A*	*B*	*C*	*D*
Number of votes	40%	60%		

A is eliminated, and *B* is declared the winner.
Winner: *B*. Second place: *A*. Third place: *C*. Last place: *D*.

   (d)   *A* versus *B*: 40% of the votes to 60% of the votes (*B* wins). *B* gets 1 point.
         A versus *C*: 75% of the votes to 25% of the votes (*A* wins). *A* gets 1 point.
         A versus *D*: 55% of the votes to 45% of the votes (*A* wins). *A* gets 1 point.
         B versus *C*: 75% of the votes to 25% of the votes (*B* wins). *B* gets 1 point.
         B versus *D*: 60% of the votes to 40% of the votes (*B* wins). *B* gets 1 point.
         C versus *D*: 25% of the votes to 75% of the votes (*D* wins). *D* gets 1 point.
         The final tally is 2 points for *A*, 3 points for *B*, 0 points for *C*, and 1 point for *D*.
         Winner: *B*. Second place: *A*. Third place: *D*. Last place: *C*.

45.   Step 1: From Example 1.5 we know that the winner using the Borda count method is *B* with 106 points.
     Step 2: Removing *B* gives the following preference schedule.

Number of voters	14	10	8	4	1
1st choice	*A*	*C*	*D*	*D*	*C*
2nd choice	*C*	*D*	*C*	*C*	*D*
3rd choice	*D*	*A*	*A*	*A*	*A*

This preference schedule can be consolidated (simplified) to produce the following schedule.

Number of voters	14	11	12
1st choice	A	C	D
2nd choice	C	D	C
3rd choice	D	A	A

A has $3 \times 14 + 2 \times 0 + 1 \times (11+12) = 65$ points;

C has $3 \times 11 + 2 \times (14+12) + 1 \times 0 = 85$ points;

D has $3 \times 12 + 2 \times 11 + 1 \times 14 = 72$ points;

In this schedule the winner using the Borda count method is C, with 85 points. Thus, second place goes to C.

Step 3: Removing C gives the following preference schedule.

Number of voters	14	11	12
1st choice	A	D	D
2nd choice	D	A	A

A has $2 \times 14 + 1 \times (11+12) = 51$ points;

D has $2 \times (11+12) + 1 \times 14 = 60$ points;

In this schedule the winner using the Borda count method is D, with 60 points. Thus, third place goes to D, and last place goes to A.

47. **(a)** Step 1: From Exercise 41(a) we know that the winner, using plurality, is A with 12 first-place votes.
Step 2: Removing A gives the following preference schedule.

Number of voters	4	1	9	8	5
1st choice	C	B	C	B	D
2nd choice	B	D	D	D	C
3rd choice	D	C	B	C	B

In this schedule the winner using plurality is C, with 13 first-place votes. Thus, second place goes to C.

Step 3: Removing C gives the following preference schedule.

Number of voters	4	1	9	8	5
1st choice	B	B	D	B	D
2nd choice	D	D	B	D	B

In this schedule the winner using plurality is D, with 14 first-place votes. Thus, third place goes to D and last place goes to B.

(b) Step 1: From Exercise 41(b) we know that the winner using the Borda count method is A with 74 points.
Step 2: Removing A gives the following preference schedule.

Number of voters	4	1	9	8	5
1st choice	C	B	C	B	D
2nd choice	B	D	D	D	C
3rd choice	D	C	B	C	B

B has $3 \times (1+8) + 2 \times 4 + 1 \times (9+5) = 49$ points;

C has $3 \times (4+9) + 2 \times 5 + 1 \times (1+8) = 58$ points;

D has $3 \times 5 + 2 \times (1+9+8) + 1 \times 4 = 55$ points;

In this schedule the winner using the Borda count method is *C*, with 58 points. Thus, second place goes to *C*.

Step 3: Removing *C* gives the following preference schedule.

Number of voters	4	1	9	8	5
1st choice	*B*	*B*	*D*	*B*	*D*
2nd choice	*D*	*D*	*B*	*D*	*B*

B has $2 \times 13 + 1 \times 14 = 40$ points;

D has $2 \times 14 + 1 \times 13 = 41$ points;

In this schedule the winner using the Borda count method is *D*, with 41 points. Thus, third place goes to *D*, and last place goes to *B*.

(c)   Step 1: From Exercise 41(c) we know that the winner using plurality-with-elimination is *C*.

Step 2: Removing *C* gives the following preference schedule.

Number of voters	4	1	9	8	5
1st choice	*A*	*B*	*D*	*A*	*D*
2nd choice	*B*	*A*	*A*	*B*	*B*
3rd choice	*D*	*D*	*B*	*D*	*A*

In this schedule the winner using plurality-with-elimination is *D* because *D* has a majority (14) of the first-place votes. Thus, second place goes to *D*.

Step 3: Removing *D* gives the following preference schedule.

Number of voters	4	1	9	6	5
1st choice	*A*	*B*	*A*	*A*	*B*
2nd choice	*B*	*A*	*B*	*B*	*A*

In this schedule the winner using plurality-with-elimination is *A* because *A* has a majority (21) of the first-place votes. Thus, third place goes to *A*, and last place goes to *B*.

(d)   Step 1: From Exercise 41(d) we know that the winner using pairwise comparisons is *D*.

Step 2: Removing *D* gives the following preference schedule.

Number of voters	4	1	9	8	5
1st choice	*A*	*B*	*C*	*A*	*C*
2nd choice	*C*	*A*	*A*	*B*	*B*
3rd choice	*B*	*C*	*B*	*C*	*A*

A versus *B*: 21 votes to 6 votes (*A* wins). *A* gets 1 point.

A versus *C*: 13 votes to 14 votes (*C* wins). *C* gets 1 point.

B versus *C*: 9 votes to 18 votes (*C* wins). *C* gets 1 point.

The final tally is 1 point for *A*, 0 points for *B*, and 2 points for *C*. In this schedule the winner using pairwise comparisons is *C*, with 2 points. Thus, second place goes to *C*.

Step 3: Removing *C* gives the following preference schedule.

Number of voters	4	1	9	8	5
1st choice	*A*	*B*	*A*	*A*	*B*
2nd choice	*B*	*A*	*B*	*B*	*A*

A versus *B*: 21 votes to 6 votes (*A* wins). *A* gets 1 point. The final tally is 1 point for *A*, 0 points for *B*. In this schedule the winner using pairwise comparisons is *A*, with 1 point. Thus, third place goes to *A*, and last place goes to *B*.

49. (a) Step 1: From Exercise 43(a) we know that the winner, using plurality, is *A* with 40% of the first-place votes.

Step 2: Removing *A* gives the following preference schedule.

Percentage of voters	40%	25%	20%	15%
1st choice	*D*	*C*	*B*	*B*
2nd choice	*B*	*B*	*D*	*D*
3rd choice	*C*	*D*	*C*	*C*

In this schedule the winner using plurality is *D*, with 40% of the first-place votes. Thus, second place goes to *D*.

Step 3: Removing *D* gives the following preference schedule.

Percentage of voters	40%	25%	20%	15%
1st choice	*B*	*C*	*B*	*B*
2nd choice	*C*	*B*	*C*	*C*

In this schedule the winner using plurality is *B*, with 75% of the first-place votes. Thus, third place goes to *B* and last place goes to *C*.

(b) Step 1: From Exercise 43(b) we know that the winner using the Borda count method is *B* with 295 Borda points.

Step 2: Removing *B* gives the following preference schedule.

Percentage of voters	40%	25%	20%	15%
1st choice	*A*	*C*	*D*	*A*
2nd choice	*D*	*D*	*A*	*D*
3rd choice	*C*	*A*	*C*	*C*

A has $3 \times (40 + 15) + 2 \times 20 + 1 \times 25 = 230$ points;

C has $3 \times 25 + 2 \times 0 + 1 \times (40 + 20 + 15) = 150$ points;

D has $3 \times 20 + 2 \times (40 + 25 + 15) + 1 \times 0 = 220$ points;

In this schedule the winner using the Borda count method is *A*, with 230 points. Thus, second place goes to *A*.

Step 3: Removing *A* gives the following preference schedule.

Percentage of voters	40%	25%	20%	15%
1st choice	*D*	*C*	*D*	*D*
2nd choice	*C*	*D*	*C*	*C*

C has $2 \times 25 + 1 \times (40 + 20 + 15) = 125$ points;

D has $2 \times (40 + 20 + 15) + 1 \times 25 = 175$ points;

In this schedule the winner using the Borda count method is *D*, with 175 points. Thus, third place goes to *D*, and last place goes to *C*.

(c) Step 1: From Exercise 43(c) we know that the winner using plurality-with-elimination is *B*.
Step 2: Removing *B* gives the following preference schedule.

Percentage of voters	40%	25%	20%	15%
1st choice	*A*	*C*	*D*	*A*
2nd choice	*D*	*D*	*A*	*D*
3rd choice	*C*	*A*	*C*	*C*

In this schedule the winner using plurality-with-elimination is *A* because *A* has a majority (55%) of the first-place votes. Thus, second place goes to *A*.
Step 3: Removing *A* gives the following preference schedule.

Percentage of voters	40%	25%	20%	15%
1st choice	*D*	*C*	*D*	*D*
2nd choice	*C*	*D*	*C*	*C*

In this schedule the winner using plurality-with-elimination is *D* because *D* has a majority (75%) of the first-place votes. Thus, third place goes to *D*, and last place goes to *C*.

(d) Step 1: From Exercise 43(d) we know that the winner using pairwise comparisons is *B*.
Step 2: Removing *B* gives the following preference schedule.

Percentage of voters	40%	25%	20%	15%
1st choice	*A*	*C*	*D*	*A*
2nd choice	*D*	*D*	*A*	*D*
3rd choice	*C*	*A*	*C*	*C*

A versus *C*: 75% of the votes to 25% of the votes (*A* wins). *A* gets 1 point.
A versus *D*: 55% of the votes to 45% of the votes (*A* wins). *A* gets 1 point.
C versus *D*: 25% of the votes to 75% of the votes (*D* wins). *D* gets 1 point.
The final tally is 2 points for *A*, 0 points for *C*, and 1 point for *D*. In this schedule the winner using pairwise comparisons is *A*, with 2 points. Thus, second place goes to *A*.

Step 3: Removing *A* gives the following preference schedule.

Percentage of voters	40%	25%	20%	15%
1st choice	*D*	*C*	*D*	*D*
2nd choice	*C*	*D*	*C*	*C*

C versus *D*: 25% of the votes to 75% of the votes (*D* wins). *D* gets 1 point. The final tally is 0 points for *C*, 1 point for *D*. In this schedule the winner using pairwise comparisons is *D*, with 1 point. Thus, third place goes to *D*, and last place goes to *C*.

G. Miscellaneous

51. $\dfrac{500 \times 501}{2} = 125{,}250$

53. $1+2+3+\ldots+3218+3219+3220 = \dfrac{3220\times3221}{2} = 5,185,810$

Combining this with the result from Exercise 51, we get
$501+502+503+\ldots+3218+3219+3220 = 5,185,810-125,250 = 5,060,560.$

55. (a) $1+2+3+\ldots+12+13+14 = \dfrac{14\times15}{2} = 105$

(b) 105 minutes = 1 hour and 45 minutes

57. Each column corresponds to a unique ordering of candidates *A*, *B*, and *C*. There are 3 choices as to which candidate is listed first, 2 choices for which candidate is listed second, and 1 choice as to which candidate is listed last. This leads to $3\times2\times1=6$ ways that the candidates can be ordered on the ballots. (These are *ABC*, *ACB*, *BAC*, *BCA*, *CAB*, *CBA*.)

59. (a) *A* has a majority of the first-place votes (7), so *A* is the Condorcet candidate.

(b) *A* has $4\times7+3\times2+2\times0+1\times4 = 38$ points;
B has $4\times4+3\times7+2\times0+1\times2 = 39$ points;
C has $4\times0+3\times0+2\times(7+4+2)+1\times0 = 26$ points;
D has $4\times2+3\times4+2\times0+1\times7 = 27$ points;
The winner is *B*.

(c) Removing *C* gives the following preference schedule.

Number of voters	7	4	2
1st choice	A	B	D
2nd choice	B	D	A
3rd choice	D	A	B

A has $3\times7+2\times2+1\times4 = 29$ points;
B has $3\times4+2\times7+1\times2 = 28$ points;
D has $3\times2+2\times4+1\times7 = 21$ points;
The winner is *A*.

(d) Based on (a) and (b), the Condorcet criterion and the majority criterion are violated. Based on (b) and (c), the independence of irrelevant alternatives criterion is violated. Furthermore, based on (b), the Borda count also violates the majority criterion since *A* has a majority of the first-place votes but does not win the election.

JOGGING

61. Suppose the two candidates are *A* and *B* and that *A* gets *a* first-place votes and *B* gets *b* first-place votes and suppose that $a > b$. Then *A* has a majority of the votes and the preference schedule is

Number of voters	a	b
1st choice	A	B
2nd choice	B	A

It is clear that candidate *A* wins the election under the plurality method, the plurality-with-elimination method, and the method of pairwise comparisons. Under the Borda count method, *A* gets $2a + b$ points while *B* gets $2b + a$ points. Since $a > b$, $2a + b > 2b + a$ and so again *A* wins the election.

63. If *X* is the winner of an election using the plurality method and, in a reelection, the only changes in the ballots are changes that only favor *X*, then no candidate other than *X* can increase his/her first-place votes and so *X* is still the winner of the election.

65. If *X* is the winner of an election using the method of pairwise comparisons and, in a reelection, the only changes in the ballots are changes that favor *X* and only favor *X*, then candidate *X* will still win every pairwise comparison that he/she won in the original election and possibly even some new ones — while no other candidate will win any new pairwise comparisons (since there were no changes favorable to any other candidate). That is, if *X* has *A* wins in the original election and *A'* wins in the reelection, then $A' \geq A$. Similarly, if *Y* has *B* wins in the original election and *B'* wins in the reelection, then $B' \leq B$. So, we have $A' - A \geq 0 \geq B' - B$ and *X* will gain more points than any other candidate will gain and hence will remain the winner of the election.

67. (a) Suppose a candidate, *C*, gets v_1 first-place votes, v_2 second-place votes, v_3 third-place votes, ..., v_N *N*th-place votes. For this candidate $p = v_1 N + v_2(N-1) + v_3(N-2) + ... + v_{N-1} \cdot 2 + v_N \cdot 1$,
    and $r = v_1 \cdot 1 + v_2 \cdot 2 + v_3 \cdot 3 + ... + v_{N-1}(N-1) + v_N N$.
    So, $p + r = v_1(N+1) + v_2(N+1) + v_3(N+1) + ... + v_{N-1}(N+1) + v_N(N+1)$
    $$= (v_1 + v_2 + ... + v_N)(N+1)$$
    $$= k(N+1)$$

    (b) Suppose candidates C_1 and C_2 receive p_1 and p_2 points respectively using the Borda count as originally described in the chapter and r_1 and r_2 points under the variation described in this exercise. Then if $p_1 < p_2$, we have $-p_1 > -p_2$ and so $k(N+1) - p_1 > k(N+1) - p_2$ which implies [using part (a)], $r_1 > r_2$. Consequently the relative ranking of the candidates is not changed.

69. (a) This follows since $1610 + 1540 + 1530 = 65 \times (25 + 24 + 23)$. That is, these three teams combined received as many first, second, and third-place votes as possible.

    (b) USC: Let x = the number of second-place votes. Then $1610 = 25 \times 52 + 24 \times x + 23 \times (65 - 52 - x)$
    and so $x = 11$. This leaves $65 - 52 - 11 = 2$ third-place votes.
    Oklahoma: Let y = the number of second-place votes. Then $1540 = 25 \times 7 + 24 \times y + 23 \times (65 - 7 - y)$
    and so $y = 31$. This leaves $65 - 7 - 31 = 27$ third-place votes.
    Auburn: Let z = the number of second-place votes. Then $1530 = 25 \times 6 + 24 \times z + 23 \times (65 - 6 - z)$
    and so $z = 23$. This leaves $65 - 6 - 23 = 36$ third-place votes.

71. By looking at Dwayne Wade's vote totals, it is clear that 1 point is awarded for each third-place vote. Let x = points awarded for each second-place vote. Then, $3x + 108 = 117$ gives $x = 3$.
    Next, let y = points awarded for each third-place vote. Looking at LeBron James' votes, we see that $78y + 39 \times 3 + 1 \times 1 = 508$. Solving this equation gives $y = 5$.
    5 points for each first-place vote; 3 points for each second-place vote; 1 point for each third-place vote.

73. (a) C

A is eliminated first, D is eliminated next, and then C beats B.

(b) A is a Condorcet candidate but is eliminated in the first round.

Number of voters	10	6	6	3	3
1st choice	B	A	A	D	C
2nd choice	C	B	C	A	A
3rd choice	D	D	B	C	B
4th choice	A	C	D	B	D

(c) B wins under the Coombs method. However, if 8 voters move B from their 3rd choice to their 2nd choice, then C wins.

Number of voters	10	8	7	4
1st choice	B	C	C	A
2nd choice	A	A	B	B
3rd choice	C	B	A	C

Chapter 2

WALKING

A. Weighted Voting Systems

1. (a) There are 6 players.

(b) $7 + 4 + 3 + 3 + 2 + 1 = 20$ votes.

(c) 4

(d) $\dfrac{13}{20} = 0.65$

65%

3. (a) The quota must be more than half of the total votes. This system has $10 + 6 + 5 + 4 + 2 = 27$ total votes. $\dfrac{1}{2} \times 27 = 13.5$, so the smallest value q can take is 14.

(b) The largest value q can take is 27, the total number of votes.

(c) $\dfrac{2}{3} \times 27 = 18$, so the value of the quota q would be 18.

(d) The value of the quota q would be strictly larger than 18. That is, 19.

5. To determine the number of votes each player has, let P_4 have 1 vote. Then P_3 has $2 \times 1 = 2$ votes; P_2 has $2 \times 2 = 4$ votes; P_1 has $2 \times 4 = 8$ votes. The parts of the problem then require determining the quotas for various scenarios when there are $1 + 2 + 4 + 8 = 15$ total votes.

(a) $\dfrac{2}{3} \times 15 = 10$

[10: 8, 4, 2, 1]

(b) [11: 8, 4, 2, 1]

(c) $0.80 \times 15 = 12$

[12: 8, 4, 2, 1]

(d) [13: 8, 4, 2, 1]

7. (a) There is no dictator; P_1 and P_2 have veto power; P_3 is a dummy.

(b) P_1 is a dictator; P_2 and P_3 are dummies.

(c) There is no dictator, no one has veto power, and no one is a dummy.

9. (a) There is no dictator; P_1 and P_2 have veto power; P_5 is a dummy.

   (b) P_1 is a dictator; P_2, P_3, and P_4 are dummies.

   (c) There is no dictator; P_1 and P_2 have veto power; P_3 and P_4 are dummies.

   (d) There is no dictator; all 4 players have veto power.

B.   Banzhaf Power

11. (a) $6 + 4 = 10$

   (b) $\{P_1, P_2\}$, $\{P_1, P_3\}$, $\{P_1, P_2, P_3\}$, $\{P_1, P_2, P_4\}$, $\{P_1, P_3, P_4\}$, $\{P_2, P_3, P_4\}$, $\{P_1, P_2, P_3, P_4\}$

   (c) P_1 only

   (d)

Winning Coalitions	Critical players
$\{P_1, P_2\}$	P_1, P_2
$\{P_1, P_3\}$	P_1, P_3
$\{P_1, P_2, P_3\}$	P_1
$\{P_1, P_2, P_4\}$	P_1, P_2
$\{P_1, P_3, P_4\}$	P_1, P_3
$\{P_2, P_3, P_4\}$	P_2, P_3, P_4
$\{P_1, P_2, P_3, P_4\}$	none

P_1 is critical 5 times; P_2 is critical 3 times; P_3 is critical 3 times; P_4 is critical one time. The total number of times the players are critical is $5 + 3 + 3 + 1 = 12$. The Banzhaf power distribution is $P_1 : \dfrac{5}{12} = 41\dfrac{2}{3}\%$; $P_2 : \dfrac{3}{12} = 25\%$; $P_3 : \dfrac{3}{12} = 25\%$; $P_4 : \dfrac{1}{12} = 8\dfrac{1}{3}\%$.

13. (a)

Winning Coalitions	Critical players
$\{P_1, P_2\}$	P_1, P_2
$\{P_1, P_3\}$	P_1, P_3
$\{P_1, P_2, P_3\}$	P_1

P_1 is critical 3 times; P_2 is critical 1 time; P_3 is critical 1 time. The total number of times players are critical is $3 + 1 + 1 = 5$. The Banzhaf power distribution is $P_1 : \dfrac{3}{5} = 60\%$;

$P_2 : \dfrac{1}{5} = 20\%$; $P_3 : \dfrac{1}{5} = 20\%$.

(b)

Winning Coalitions	Critical players
$\{P_1, P_2\}$	P_1, P_2
$\{P_1, P_3\}$	P_1, P_3
$\{P_1, P_2, P_3\}$	P_1

P_1 is critical 3 times; P_2 is critical 1 time; P_3 is critical 1 time. The total number of times the

players are critical is $3 + 1 + 1 = 5$. The Banzhaf power distribution is $P_1 : \dfrac{3}{5} = 60\%$; $\quad P_2 : \dfrac{1}{5} = 20\%$;

$P_3 : \dfrac{1}{5} = 20\%$. The answers to (a) and (b) are the same.

15. (a)

Winning Coalitions	Critical players
$\{P_1, P_2, P_3\}$	P_1, P_2, P_3
$\{P_1, P_2, P_4\}$	P_1, P_2, P_4
$\{P_1, P_2, P_5\}$	P_1, P_2, P_5
$\{P_1, P_3, P_4\}$	P_1, P_3, P_4
$\{P_1, P_2, P_3, P_4\}$	P_1
$\{P_1, P_2, P_3, P_5\}$	P_1, P_2
$\{P_1, P_2, P_4, P_5\}$	P_1, P_2
$\{P_1, P_3, P_4, P_5\}$	P_1, P_3, P_4
$\{P_2, P_3, P_4, P_5\}$	P_2, P_3, P_4, P_5
$\{P_1, P_2, P_3, P_4, P_5\}$	none

P_1 is critical 8 times; P_2 is critical 6 times; P_3 is critical 4 times; P_4 is critical 4 times; P_5 is critical 2 times. The total number of times the players are critical is $8 + 6 + 4 + 4 + 2 = 24$. The Banzhaf power distribution is $P_1 : \dfrac{8}{24} = 33\dfrac{1}{3}\%$; $\quad P_2 : \dfrac{6}{24} = 25\%$; $\quad P_3 : \dfrac{4}{24} = 16\dfrac{2}{3}\%$; $\quad P_4 : \dfrac{4}{24} = 16\dfrac{2}{3}\%$;

$P_5 : \dfrac{2}{24} = 8\dfrac{1}{3}\%$.

(b) The quota is one more than in (a), so some winning coalitions may now be losing coalitions. For the ones that are still winning, any players that were critical in (a) will still be critical, and there may be additional critical players. A quick check shows that $\{P_1, P_2, P_5\}$, $\{P_1, P_3, P_4\}$, and $\{P_2, P_3, P_4, P_5\}$ are now losing coalitions (they all have exactly 10 votes).

Winning Coalitions	Critical players
$\{P_1, P_2, P_3\}$	P_1, P_2, P_3
$\{P_1, P_2, P_4\}$	P_1, P_2, P_4
$\{P_1, P_2, P_3, P_4\}$	P_1, P_2
$\{P_1, P_2, P_3, P_5\}$	P_1, P_2, P_3
$\{P_1, P_2, P_4, P_5\}$	P_1, P_2, P_4
$\{P_1, P_3, P_4, P_5\}$	P_1, P_3, P_4, P_5
$\{P_1, P_2, P_3, P_4, P_5\}$	P_1

P_1 is critical 7 times; P_2 is critical 5 times; P_3 is critical 3 times; P_4 is critical 3 times; P_5 is critical 1 time. The total number of times the players are critical is $7 + 5 + 3 + 3 + 1 = 19$. The Banzhaf power distribution is $P_1 : \frac{7}{19} \approx 36\frac{13}{19}\%$; $P_2 : \frac{5}{19} \approx 26\frac{12}{19}\%$; $P_3 : \frac{3}{19} \approx 15\frac{11}{19}\%$; $P_4 : \frac{3}{19} \approx 15\frac{11}{19}\%$;

$P_5 : \frac{1}{19} \approx 5\%$.

17. **(a)** P_1 is a dictator and the other players are dummies. Thus P_1 is the only critical player in each winning coalition. The Banzhaf power distribution is $P_1 : 1$; $P_2 : 0$; $P_3 : 0$; $P_4 : 0$.

(b)

Winning Coalitions	Critical players
$\{P_1, P_2\}$	P_1, P_2
$\{P_1, P_3\}$	P_1, P_3
$\{P_1, P_4\}$	P_1, P_4
$\{P_1, P_2, P_3\}$	P_1
$\{P_1, P_2, P_4\}$	P_1
$\{P_1, P_3, P_4\}$	P_1
$\{P_1, P_2, P_3, P_4\}$	P_1

P_1 is critical 7 times; P_2 is critical 1 time; P_3 is critical 1 time; P_4 is critical 1 time. The total number of times all players are critical is $7 + 1 + 1 + 1 = 10$. The Banzhaf power distribution is $P_1 : \frac{7}{10} = 70\%$; $P_2 : \frac{1}{10} = 10\%$; $P_3 : \frac{1}{10} = 10\%$; $P_4 : \frac{1}{10} = 10\%$.

(c) This situation is like (b) with the following exceptions: $\{P_1, P_4\}$ is now a losing coalition; P_1 and P_2 are both critical in $\{P_1, P_2, P_4\}$; P_1 and P_3 are both critical in $\{P_1, P_3, P_4\}$. Now P_1 is critical 6 times; P_2 is critical 2 times; P_3 is critical 2 times; P_4 is never critical. The total number of times all players are critical is $6 + 2 + 2 = 10$. The Banzhaf power distribution is $P_1 : \frac{6}{10} = 60\%$;

$P_2 : \frac{2}{10} = 20\%$; $P_3 : \frac{2}{10} = 20\%$; $P_4 = 0$.

(d)

Winning Coalitions	Critical players
$\{P_1, P_2\}$	P_1, P_2
$\{P_1, P_2, P_3\}$	P_1, P_2
$\{P_1, P_2, P_4\}$	P_1, P_2
$\{P_1, P_2, P_3, P_4\}$	P_1, P_2

P_1 and P_2 are each critical in every winning coalition, and P_3 and P_4 are never critical. The Banzhaf power distribution is $P_1 : \frac{1}{2} = 50\%$; $P_2 : \frac{1}{2} = 50\%$; $P_3 : 0$; $P_4 : 0$.

(e)

Winning Coalitions	Critical players
$\{P_1, P_2, P_3\}$	P_1, P_2, P_3
$\{P_1, P_2, P_3, P_4\}$	P_1, P_2, P_3

P_1, P_2, and P_3 are each critical in every winning coalition. P_4 is never critical. The Banzhaf power distribution is $P_1 : \frac{1}{3} = 33\frac{1}{3}\%$; $P_2 : \frac{1}{3} = 33\frac{1}{3}\%$; $P_3 : \frac{1}{3} = 33\frac{1}{3}\%$; $P_4 : 0$.

19. D is never a critical player, and the other three have equal power. Thus, the Banzhaf power distribution is

$A : \frac{1}{3} = 33\frac{1}{3}\%$; $B : \frac{1}{3} = 33\frac{1}{3}\%$; $C : \frac{1}{3} = 33\frac{1}{3}\%$; $D: 0$.

21. (a) $\{P_1, P_2\}$, $\{P_1, P_3\}$, $\{P_2, P_3\}$, $\{P_1, P_2, P_3\}$

(b) $\{P_1, P_2, P_4\}$, $\{P_1, P_3, P_4\}$, $\{P_2, P_3, P_4\}$, $\{P_1, P_2, P_3, P_4\}$, $\{P_1, P_2, P_4, P_5\}$, $\{P_1, P_2, P_4, P_6\}$, $\{P_1, P_3, P_4, P_5\}$, $\{P_1, P_3, P_4, P_6\}$, $\{P_2, P_3, P_4, P_5\}$, $\{P_2, P_3, P_4, P_6\}$, $\{P_1, P_2, P_3, P_4, P_5\}$, $\{P_1, P_2, P_3, P_4, P_6\}$, $\{P_1, P_2, P_3, P_4, P_5, P_6\}$

(c) P_4 is never a critical player since every time it is part of a winning coalition, that coalition is a winning coalition without P_4 as well.

(d) A similar argument to that used in part (c) shows that P_5 and P_6 are also dummies. One could also argue that any player with fewer votes than P_4, a dummy, will also be a dummy. So, P_4, P_5, and P_6 will never be critical -- they all have zero power.

The only winning coalitions with only 2 players are $\{P_1, P_2\}$, $\{P_1, P_3\}$, and $\{P_2, P_3\}$; and both players are critical in each of those coalitions. All other winning coalitions consist of one of these coalitions plus additional players, and the only critical players will be the ones from the two-player coalition. So P_1, P_2, and P_3 will be critical in every winning coalition they are in, and they will all be in the same number of winning coalitions, so they all have the same power. Thus, the Banzhaf power distribution is $P_1 : \frac{1}{3} = 33\frac{1}{3}\%; \quad P_2 : \frac{1}{3} = 33\frac{1}{3}\%; \quad P_3 : \frac{1}{3} = 33\frac{1}{3}\%; \quad P_4 : 0; \quad P_5 : 0; \quad P_6 : 0.$

C.   Shapley-Shubik Power

23. (a) There are 3! = 6 sequential coalitions of the three players.

$< P_1, \underline{P_2}, P_3 >, < P_1, \underline{P_3}, P_2 >, < P_2, \underline{P_1}, P_3 >, < P_2, P_3, \underline{P_1} >, < P_3, \underline{P_1}, P_2 >, < P_3, P_2, \underline{P_1} >$

(b) P_1 is pivotal 4 times; P_2 is pivotal 1 time; P_3 is pivotal 1 time. The Shapley-Shubik power distribution is $P_1 : \frac{4}{6} = 66\frac{2}{3}\%; \quad P_2 : \frac{1}{6} = 16\frac{2}{3}\%; \quad P_3 : \frac{1}{6} = 16\frac{2}{3}\%.$

25. There are 4! = 24 sequential coalitions of the four players. The pivotal player in each coalition is underlined.

$< P_1, \underline{P_2}, P_3, P_4 >, < P_1, \underline{P_2}, P_4, P_3 >, < P_1, P_3, \underline{P_2}, P_4 >, < P_1, P_3, \underline{P_4}, P_2 >,$

$< P_1, P_4, \underline{P_2}, P_3 >, < P_1, P_4, \underline{P_3}, P_2 >, < P_2, \underline{P_1}, P_3, P_4 >, < P_2, \underline{P_1}, P_4, P_3 >,$

$< P_2, P_3, \underline{P_1}, P_4 >, < P_2, P_3, P_4, \underline{P_1} >, < P_2, P_4, \underline{P_1}, P_3 >, < P_2, P_4, P_3, \underline{P_1} >,$

$< P_3, P_1, \underline{P_2}, P_4 >, < P_3, P_1, \underline{P_4}, P_2 >, < P_3, P_2, \underline{P_1}, P_4 >, < P_3, P_2, P_4, \underline{P_1} >,$

$< P_3, P_4, \underline{P_1}, P_2 >, < P_3, P_4, P_2, \underline{P_1} >, < P_4, P_1, \underline{P_2}, P_3 >, < P_4, P_1, \underline{P_3}, P_2 >,$

$< P_4, P_2, \underline{P_1}, P_3 >, < P_4, P_2, P_3, \underline{P_1} >, < P_4, P_3, \underline{P_1}, P_2 >, < P_4, P_3, P_2, \underline{P_1} >$

P_1 is pivotal 14 times; P_2 is pivotal 6 times; P_3 is pivotal 2 times; P_4 is pivotal 2 times.

The Shapley-Shubik power distribution is $P_1 : \frac{14}{24} = 50\frac{1}{3}\%; \quad P_2 : \frac{6}{24} = 25\%; \quad P_3 : \frac{2}{24} = 8\frac{1}{3}\%;$

$P_4 : \frac{2}{24} = 8\frac{1}{3}\%.$

27. There are 3! = 6 sequential coalitions of the three players. The pivotal player in each coalition is underlined.

(a) P_1 is a dictator and will be the pivotal player in every coalition. The Shapley-Shubik power distribution is $P_1 : 1; P_2 : 0; P_3 : 0.$

(b) $< P_1, \underline{P_2}, P_3 >, < P_1, \underline{P_3}, P_2 >, < P_2, \underline{P_1}, P_3 >, < P_2, P_3, \underline{P_1} >, < P_3, \underline{P_1}, P_2 >, < P_3, P_2, \underline{P_1} >$
P_1 is pivotal 4 times; P_2 is pivotal 1 time; P_3 is pivotal 1 time. The Shapley-Shubik power distribution is $P_1 : \frac{4}{6} = 66\frac{2}{3}\%; \quad P_2 : \frac{1}{6} = 16\frac{2}{3}\%; \quad P_3 : \frac{1}{6} = 16\frac{2}{3}\%.$

(c) This is effectively the same system as in (b), so the Shapley-Shubik power distribution is $P_1 : \frac{4}{6} = 66\frac{2}{3}\%; \quad P_2 : \frac{1}{6} = 16\frac{2}{3}\%; \quad P_3 : \frac{1}{6} = 16\frac{2}{3}\%.$

(d) $< P_1, \underline{P_2}, P_3 >, < P_1, P_3, \underline{P_2} >, < P_2, \underline{P_1}, P_3 >, < P_2, P_3, \underline{P_1} >, < P_3, P_1, \underline{P_2} >, < P_3, P_2, \underline{P_1} >$

P_1 is pivotal 3 times; P_2 is pivotal 3 times; P_3 is never pivotal. The Shapley-Shubik power distribution is $P_1 : \dfrac{3}{6} = 50\%; P_2 : \dfrac{3}{6} = 50\%; P_3 : 0.$

(e) $< P_1, \underline{P_2}, P_3 >, < P_1, \underline{P_3}, P_2 >, < P_2, \underline{P_1}, P_3 >, < P_2, \underline{P_3}, P_1 >, < P_3, \underline{P_1}, P_2 >, < P_3, \underline{P_2}, P_1 >$

Each player is pivotal 2 times. The Shapley-Shubik power distribution is $P_1 : \dfrac{1}{3} = 33\dfrac{1}{3}\%;$

$P_2 : \dfrac{1}{3} = 33\dfrac{1}{3}\%; \quad P_3 : \dfrac{1}{3} = 33\dfrac{1}{3}\%.$

29. There are $3! = 6$ sequential coalitions of the three players. The pivotal player in each coalition is underlined.

(a) P_1 is a dictator and will be the pivotal player in every coalition. The Shapley-Shubik power distribution is $P_1 : 1; P_2 : 0; P_3 : 0.$

(b) $< P_1, \underline{P_2}, P_3 >, < P_1, \underline{P_3}, P_2 >, < P_2, \underline{P_1}, P_3 >, < P_2, P_3, \underline{P_1} >, < P_3, \underline{P_1}, P_2 >, < P_3, P_2, \underline{P_1} >$

P_1 is pivotal 4 times; P_2 is pivotal 1 time; P_3 is pivotal 1 time. The Shapley-Shubik power distribution is $P_1 : \dfrac{4}{6} = 66\dfrac{2}{3}\%; P_2 : \dfrac{1}{6} = 16\dfrac{2}{3}\%; P_3 : \dfrac{1}{6} = 16\dfrac{2}{3}\%.$

(c) $< P_1, \underline{P_2}, P_3 >, < P_1, P_3, \underline{P_2} >, < P_2, \underline{P_1}, P_3 >, < P_2, P_3, \underline{P_1} >, < P_3, P_1, \underline{P_2} >, < P_3, P_2, \underline{P_1} >$

P_1 and P_2 are each pivotal 3 times; P_3 is never pivotal. The Shapley-Shubik power distribution is

$P_1 : \dfrac{1}{2} = 50\%; P_2 : \dfrac{1}{2} = 50\%; P_3 : 0.$

(d) This is effectively the same system as in (c), so the Shapley-Shubik power distribution is

$P_1 : \dfrac{1}{2} = 50\%; P_2 : \dfrac{1}{2} = 50\%; P_3 : 0.$

(e) In this system, the last player in the coalition will always be the pivotal player. Since every player is last in the same number of sequential coalitions, the players all have the same power. Thus, the Shapley-Shubik power distribution is $P_1 : \dfrac{1}{3} = 33\dfrac{1}{3}\%; P_2 : \dfrac{1}{3} = 33\dfrac{1}{3}\%; P_3 : \dfrac{1}{3} = 33\dfrac{1}{3}\%.$

31. (a) There are $4! = 24$ sequential coalitions of the four players. In each coalition the pivotal player is underlined.

$< P_1, \underline{P_2}, P_3, P_4 >, < P_1, \underline{P_2}, P_4, P_3 >, < P_1, \underline{P_3}, P_2, P_4 >, < P_1, \underline{P_3}, P_4, P_2 >,$

$< P_1, P_4, \underline{P_2}, P_3 >, < P_1, P_4, \underline{P_3}, P_2 >, < P_2, \underline{P_1}, P_3, P_4 >, < P_2, \underline{P_1}, P_4, P_3 >,$

$< P_2, P_3, \underline{P_1}, P_4 >, < P_2, P_3, \underline{P_4}, P_1 >, < P_2, P_4, \underline{P_1}, P_3 >, < P_2, P_4, \underline{P_3}, P_1 >,$

$<P_3, \underline{P_1}, P_2, P_4>, <P_3, \underline{P_1}, P_4, P_2>, <P_3, P_2, \underline{P_1}, P_4>, <P_3, P_2, \underline{P_4}, P_1>,$

$<P_3, P_4, \underline{P_1}, P_2>, <P_3, P_4, \underline{P_2}, P_1>, <P_4, P_1, \underline{P_2}, P_3>, <P_4, P_1, \underline{P_3}, P_2>,$

$<P_4, P_2, \underline{P_1}, P_3>, <P_4, P_2, \underline{P_3}, P_1>, <P_4, P_3, \underline{P_1}, P_2>, <P_4, P_3, \underline{P_2}, P_1>$

P_1 is pivotal in 10 coalitions; P_2 is pivotal in 6 coalitions; P_3 is pivotal in 6 coalitions; P_4 is pivotal

in 2 coalitions. The Shapley-Shubik power distribution is $P_1 : \dfrac{10}{24} = 41\dfrac{2}{3}\%; \ P_2 : \dfrac{6}{24} = 25\%;$

$P_3 : \dfrac{6}{24} = 25\%; \ P_4 \dfrac{2}{24} = 8\dfrac{1}{3}\%.$

(b) This is the same situation as in (a) – there is essentially no difference between 41 and 49 because the players' votes are all multiples of 10. The Shapley-Shubik power distribution is thus still

$P_1 : \dfrac{5}{12} = 41\dfrac{2}{3}\%; \ P_2 : \dfrac{3}{12} = 25\%; \ P_3 : \dfrac{3}{12} = 25\%; \ P_4 : \dfrac{1}{12} = 8\dfrac{1}{3}\%.$

(c) This is also the same situation as in (a) – any time a group of players has 51 votes, they must have

60 votes. The Shapley-Shubik power distribution is thus still $P_1 : \dfrac{5}{12} = 41\dfrac{2}{3}\%; \ P_2 : \dfrac{3}{12} = 25\%;$

$P_3 : \dfrac{3}{12} = 25\%; \ P_4 : \dfrac{1}{12} = 8\dfrac{1}{3}\%.$

33. D will never be pivotal and has no power. Each of the other players will be pivotal in the same number of coalitions, so they will all have equal power. Thus, the Shapley-Shubik power distribution is

$A : \dfrac{1}{3} = 33\dfrac{1}{3}\%; \ B : \dfrac{1}{3} = 33\dfrac{1}{3}\%; \ C : \dfrac{1}{3} = 33\dfrac{1}{3}\%; \ D : 0.$

D.   Miscellaneous

35. (a) $13! = 6,227,020,800$

(b) $18! = 6,402,373,705,728,000 \approx 6.402374 \times 10^{15}$

(c) $25! = 15,511,210,043,330,985,984,000,000 \approx 1.551121 \times 10^{25}$

(d) There are 25! sequential coalitions of 25 players.

25! sequential coaltions $\times \dfrac{1 \text{ second}}{1,000,000 \text{ sequential coalitions}} \times \dfrac{1 \text{ hour}}{3600 \text{ seconds}} \times \dfrac{1 \text{ day}}{24 \text{ hours}} \times \dfrac{1 \text{ year}}{365 \text{ days}}$

$\approx 4.92 \times 10^{11}$ years

This is roughly 500 billion years.

37. (a) $10! = 10 \times 9 \times 8 \times \ldots \times 3 \times 2 \times 1$

$= 10 \times 9!$

So, $9! = \dfrac{10!}{10} = \dfrac{3,628,800}{10} = 362,880$.

(b) $11! = 11 \times 10 \times 9 \times \ldots \times 3 \times 2 \times 1 = 11 \times 10!$

So, $\dfrac{11!}{10!} = \dfrac{11 \times 10!}{10!} = 11$.

(c) $11! = 11 \times 10 \times (9 \times \ldots \times 3 \times 2 \times 1) = 11 \times 10 \times 9!$

So, $\dfrac{11!}{9!} = \dfrac{11 \times 10 \times 9!}{9!} = 11 \times 10 = 110$.

(d) $\dfrac{9!}{6!} = \dfrac{9 \times 8 \times 7 \times 6!}{6!} = 9 \times 8 \times 7 = 504$

(e) $\dfrac{101!}{99!} = \dfrac{101 \times 100 \times 99!}{99!} = 101 \times 100 = 10,100$

39. (a) $\dfrac{9! + 11!}{10!} = \dfrac{9!}{10!} + \dfrac{11!}{10!} = \dfrac{9!}{10 \times 9!} + \dfrac{11 \times 10!}{10!} = \dfrac{1}{10} + 11 = 11.1$

(b) $\dfrac{101! + 99!}{100!} = \dfrac{101!}{100!} + \dfrac{99!}{100!} = \dfrac{101 \times 100!}{100!} + \dfrac{99!}{100 \times 99!} = 101 + \dfrac{1}{100} = 101.01$

41. (a) $2^6 - 1 = 63$ coalitions

(b) There are $2^5 - 1 = 31$ coalitions of the remaining five players $P_1, P_2, P_3, P_4,$ and P_5. These are exactly those coalitions that do not include P_6.

(c) There are $2^4 - 1 = 15$ coalitions of the remaining four players P_1, P_2, P_3, P_4. These are exactly those coalitions that do not include P_5 or P_6.

(d) $63 - 15 = 48$ coalitions include P_5 and P_6.

43. (a) $6! = 720$ sequential coalitions

(b) There are $5! = 120$ sequential coalitions of the remaining five players P_2, P_3, P_4, P_5 and P_6. These are exactly those coalitions that have P_1 as the last player.

(c) $720 - 120 = 600$ sequential coalitions do not have P_1 as the last player.

45. (a) There are $6 + 5 + 4 + 4 + 3 + 2 = 24$ votes in this system. The smallest value of the quota q is more than half of this or 13. No player has veto power when $q = 13$. The strongest player (P_1 having 6 votes) gains veto power if the sum of the votes of the other players is less than the quota q. This happens if $q > 18$. Thus, no player has veto power when $q = 13, 14, 15, 16, 17,$ or 18 (i.e., $13 \le q \le 18$).

(b) When $q = 19$, P_1 has veto power. However, when $q = 20$, P_2 will also gain veto power. So the only value of q for which P_1 is the only player with veto power is $q = 19$.

(c) Clearly, every player has veto power when $q = 24$. However, every player also has veto power when $q = 23$. Also P_6, the weakest player, does not have veto power when $q = 22$. It follows that every player has veto power when $23 \leq q \leq 24$.

(d) The only values of q for which the only winning coalition is the grand coalition are the same as those in part (c), namely $q = 23$ and $q = 24$.

(e) The answers to (c) and (d) are the same. The only winning coalition is the grand coalition if and only if every player has veto power.

JOGGING

47.

Winning Coalitions	Critical players
$\{P_1, P_2, P_3\}$	P_1, P_2, P_3
$\{P_1, P_2, P_4\}$	P_1, P_2, P_4
$\{P_1, P_2, P_3, P_4\}$	P_1, P_2
$\{P_1, P_2, P_3, P_5\}$	P_1, P_2, P_3
$\{P_1, P_2, P_4, P_5\}$	P_1, P_2, P_4
$\{P_1, P_2, P_3, P_4, P_5\}$	P_1, P_2

Finding the critical players in the above table can be done as follows. Since there are no winning two-player coalitions, every player in a three-player coalition is critical. To decide if a given player in a four-player coalition is critical, one need only look for the remaining players to appear as a winning three-player coalition. If the remaining players are a winning coalition, then that given player is not critical. Similarly in the grand (five-player) coalition, one need only look at the winning four-player coalitions to decide which players are critical. According to the table, the Banzhaf power distribution is

$$P_1 : \frac{6}{16} = 37.5\%; \quad P_2 : \frac{6}{16} = 37.5\%; \quad P_3 : \frac{2}{16} = 12.5\%; \quad P_4 : \frac{2}{16} = 12.5\%; \quad P_5 : 0.$$

49. (a) $6! = 720$ sequential coalitions

(b) Since the number of votes in the system is the same as the quota, the player P_6 must be the last (sixth) player in the sequential coalition in order to be pivotal (In fact, for any player to be pivotal, they must be last). In Exercise 43(b), we saw that there are $5! = 120$ sequential coalitions for which this happens.

(c) $\dfrac{120}{720} = \dfrac{1}{6} = 16\dfrac{2}{3}\%$

(d) Since each player must be the last player in a sequential coalition in order to be pivotal, the Shapley-Shubik power index of each player is $\frac{1}{6} = 16\frac{2}{3}\%$ (Each player is the last player in 120 of the 720 sequential coalitions.)

(e) If the quota equals the sum of all the weights, then the only way a player can be pivotal is for the player to be the last player in the sequential coalition. Since every player will be the last player in the same number of sequential coalitions, all players must have the same Shapley-Shubik power index. It follows that each of the N players has a Shapley-Shubik power index of $\frac{1}{N}$.

51. **(a)** [4: 2, 1, 1, 1] or [9: 5, 2, 2, 2] are among the possible answers.

(b) The sequential coalitions (with pivotal players underlined) are:
$< H, \underline{A_1}, A_2, A_3 >, < H, A_1, \underline{A_3}, A_2 >, < H, A_2, \underline{A_1}, A_3 >, < H, A_2, \underline{A_3}, A_1 >,$
$< H, A_3, \underline{A_1}, A_2 >, < H, A_3, \underline{A_2}, A_1 >, < A_1, H, \underline{A_2}, A_3 >, < A_1, H, \underline{A_3}, A_2 >,$
$< A_1, A_2, \underline{H}, A_3 >, < A_1, A_2, A_3, \underline{H} >, < A_1, A_3, \underline{H}, A_2 >, < A_1, A_3, A_2, \underline{H} >,$
$< A_2, H, \underline{A_1}, A_3 >, < A_2, H, \underline{A_3}, A_1 >, < A_2, A_1, \underline{H}, A_3 >, < A_2, A_1, A_3, \underline{H} >,$
$< A_2, A_3, \underline{H}, A_1 >, < A_2, A_3, A_1, \underline{H} >, < A_3, H, \underline{A_1}, A_2 >, < A_3, H, \underline{A_2}, A_1 >,$
$< A_3, A_1, \underline{H}, A_2 >, < A_3, A_1, A_2, \underline{H} >, < A_3, A_2, \underline{H}, A_1 >, < A_3, A_2, A_1, \underline{H} >.$
H is pivotal in 12 coalitions; A_1 is pivotal in 4 coalitions; A_2 is pivotal in 4 coalitions; A_3 is pivotal in 4 coalitions. The Shapley-Shubik power distribution is $H : \frac{12}{24} = \frac{1}{2}$; $A_1 : \frac{4}{24} = \frac{1}{6}$;
$A_2 : \frac{4}{24} = \frac{1}{6}$; $A_3 : \frac{4}{24} = \frac{1}{6}$.

53. **(a)** Suppose that a winning coalition that contains P is not a winning coalition without P. Then, by definition, P would be a critical player in that coalition.

(b) Suppose that P is a member of every winning coalition. Then removing P from any winning coalition cannot result in another winning coalition. So P is critical in every winning coalition. Since there is at least one winning coalition (the grand coalition), P is not a dummy.

(c) Suppose that P is a pivotal member of some sequential coalition. Let's say that P is the kth player listed. Then, consider the k-player coalition consisting of the previous k-1 players listed in that sequential coalition and P. P would be a critical player in that coalition and is therefore not a dummy.

(d) Suppose that P is never a pivotal member in a sequential coalition. If P were ever critical in some coalition S, then a sequential coalition that lists the members of S other than P, then P, and then the voters not in S would have P as its pivotal player.

55. **(a)** The quota must be at least half of the total number of votes and not more than the total number of votes. $7 \le q \le 13$.

(b) For $q = 7$ or $q = 8$, P_1 is a dictator because $\{P_1\}$ is a winning coalition.

(c) For $q = 9$, only P_1 has veto power since P_2 and P_3 together have just 5 votes.

(d) For $10 \leq q \leq 12$, both P_1 and P_2 have veto power since no motion can pass without both of their votes. For $q = 13$, all three players have veto power.

(e) For $q = 7$ or $q = 8$, both P_2 and P_3 are dummies because P_1 is a dictator. For $10 \leq q \leq 12$, P_3 is a dummy since all winning coalitions contain $\{P_1, P_2\}$ which is itself a winning coalition.

57. (a) $[24: 14, 8, 6, 4]$ is just $[12: 7, 4, 3, 2]$ with each value multiplied by 2. Both have Banzhaf power distribution $P_1 : \dfrac{2}{5} = 40\%$; $P_2 : \dfrac{1}{5} = 20\%$; $P_3 : \dfrac{1}{5} = 20\%$; $P_4 : \dfrac{1}{5} = 20\%$.

(b) In the weighted voting system $[q : w_1, w_2,\ldots, w_N]$, if P_k is critical in a coalition then the sum of the weights of all the players in that coalition (including P_k) is at least q, but the sum of the weights of all the players in the coalition except P_k is less than q. Consequently, if the weights of all the players in that coalition are multiplied by $c > 0$ ($c = 0$ would make no sense), then the sum of the weights of all the players in the coalition (including P_k) is at least cq but the sum of the weights of all the players in the coalition except P_k is less than cq. Therefore P_k is critical in the same coalition in the weighted voting system $[cq : cw_1, cw_2, \ldots, cw_N]$. Since the critical players are the same in both weighted voting systems, the Banzhaf power distributions will be the same.

59. (a) There are $5! = 120$ ways that P_1 can be the last player in a sequential coalition (this is the number of sequential coalitions consisting of the other 5 players).

(b) There are $5! = 120$ ways that P_1 can be the fifth player in a sequential coalition (the other players can be ordered this many ways).

(c) P_1 is not pivotal as the fourth, third, second, or first player in a sequential coalition. It follows that P_1 is pivotal in $120 + 120 = 240$ of the 720 sequential coalitions. So, the Shapley-Shubik power index of P_1 is $\dfrac{240}{720} = \dfrac{1}{3} = 33\dfrac{1}{3}\%$.

(d) Since all of the other players in the system have same weight, it follows that they will share the remaining power equally. That is, these five players share $1 - \dfrac{240}{720} = \dfrac{480}{720}$ of the power. Since $\dfrac{1}{5}\left(\dfrac{480}{720}\right) = \dfrac{96}{720} = \dfrac{2}{15}$, it follows that the Shapley-Shibik power distribution is given by

$$P_1 : \dfrac{1}{3} = 33\dfrac{1}{3}\%; \ P_2 : \dfrac{2}{15} = 13\dfrac{1}{3}\%; \ P_3 : \dfrac{2}{15} = 13\dfrac{1}{3}\%; \ P_4 : \dfrac{2}{15} = 13\dfrac{1}{3}\%; \ P_5 : \dfrac{2}{15} = 13\dfrac{1}{3}\%; \ P_6 : \dfrac{2}{15} = 13\dfrac{1}{3}\%.$$

61. You should buy your vote from P_1. The following table explains why.

Buying a vote from	Resulting weighted voting system	Resulting Banzhaf power distribution	Your power
P_1	$[6: 3, 2, 2, 2, 2]$	$P_1 : \frac{1}{5}; P_2 : \frac{1}{5}; P_3 : \frac{1}{5}; P_4 : \frac{1}{5}; P_5 : \frac{1}{5}$	$\frac{1}{5}$
P_2	$[6: 4, 1, 2, 2, 2]$	$P_1 : \frac{1}{2}; P_2 : 0; P_3 : \frac{1}{6}; P_4 : \frac{1}{6}; P_5 : \frac{1}{6}$	$\frac{1}{6}$
P_3	$[6: 4, 2, 1, 2, 2]$	$P_1 : \frac{1}{2}; P_2 : \frac{1}{6}; P_3 : 0; P_4 : \frac{1}{6}; P_5 : \frac{1}{6}$	$\frac{1}{6}$
P_4	$[6: 4, 2, 2, 1, 2]$	$P_1 : \frac{1}{2}; P_2 : \frac{1}{6}; P_3 : \frac{1}{6}; P_4 : 0; P_5 : \frac{1}{6}$	$\frac{1}{6}$

63. (a) You should buy your vote from P_2. The following table explains why.

Buying a vote from	Resulting weighted voting system	Resulting Banzhaf power distribution	Your power
P_1	$[18: 9, 8, 6, 4, 3]$	$P_1 : \frac{4}{13}; P_2 : \frac{3}{13}; P_3 : \frac{3}{13}; P_4 : \frac{2}{13}; P_5 : \frac{1}{13}$	$\frac{1}{13}$
P_2	$[18: 10, 7, 6, 4, 3]$	$P_1 : \frac{9}{25}; P_2 : \frac{1}{5}; P_3 : \frac{1}{5}; P_4 : \frac{3}{25}; P_5 : \frac{3}{25}$	$\frac{3}{25}$
P_3	$[18: 10, 8, 5, 4, 3]$	$P_1 : \frac{5}{12}; P_2 : \frac{1}{4}; P_3 : \frac{1}{6}; P_4 : \frac{1}{12}; P_5 : \frac{1}{12}$	$\frac{1}{12}$
P_4	$[18: 10, 8, 6, 3, 3]$	$P_1 : \frac{5}{12}; P_2 : \frac{1}{4}; P_3 : \frac{1}{6}; P_4 : \frac{1}{12}; P_5 : \frac{1}{12}$	$\frac{1}{12}$

(b) You should buy 2 votes from P_2. The following table explains why.

Buying a vote from	Resulting weighted voting system	Resulting Banzhaf power distribution	Your power
P_1	$[18: 8, 8, 6, 4, 4]$	$P_1 : \frac{7}{27}; P_2 : \frac{7}{27}; P_3 : \frac{7}{27}; P_4 : \frac{1}{9}; P_5 : \frac{1}{9}$	$\frac{1}{9}$
P_2	$[18: 10, 6, 6, 4, 4]$	$P_1 : \frac{5}{13}; P_2 : \frac{2}{13}; P_3 : \frac{2}{13}; P_4 : \frac{2}{13}; P_5 : \frac{2}{13}$	$\frac{2}{13}$
P_3	$[18: 10, 8, 4, 4, 4]$	$P_1 : \frac{11}{25}; P_2 : \frac{1}{5}; P_3 : \frac{3}{25}; P_4 : \frac{3}{25}; P_5 : \frac{3}{25}$	$\frac{3}{25}$
P_4	$[18: 10, 8, 6, 2, 4]$	$P_1 : \frac{9}{25}; P_2 : \frac{7}{25}; P_3 : \frac{1}{5}; P_4 : \frac{1}{25}; P_5 : \frac{3}{25}$	$\frac{3}{25}$

(c) Buying a single vote from P_2 raises your power from $\frac{1}{25} = 4\%$ to $\frac{3}{25} = 12\%$. Buying a second vote from P_2 raises your power to $\frac{2}{13} \approx 15.4\%$. The increase in power is less with the second vote, but if you value power over money, it might still be worth it to you to buy that second vote.

65. (a) The losing coalitions are $\{P_1\}$, $\{P_2\}$, and $\{P_3\}$. The complements of these coalitions are $\{P_2, P_3\}$, $\{P_1, P_3\}$, and $\{P_1, P_2\}$ respectively, all of which are winning coalitions.

(b) The losing coalitions are $\{P_1\}, \{P_2\}, \{P_3\}, \{P_4\}, \{P_2, P_3\}, \{P_2, P_4\}$, and $\{P_3, P_4\}$. The complements of these coalitions are $\{P_2, P_3, P_4\}, \{P_1, P_3, P_4\}, \{P_1, P_2, P_4\}, \{P_1, P_2, P_3\}, \{P_1, P_4\}, \{P_1, P_3\}$, and $\{P_1, P_2\}$ respectively, all of which are winning coalitions.

(c) If P is a dictator, the losing coalitions are all the coalitions without P; the winning coalitions are all the coalitions that include P. The complement of any coalition without P (losing) is a coalition with P (winning).

(d) Take the grand coalition out of the picture for a moment. Of the remaining $2^N - 2$ coalitions, half are losing coalitions and half are winning coalitions, since each losing coalition pairs up with a winning coalition (its complement). Half of $2^N - 2$ is $2^{N-1} - 1$. In addition, we have the grand coalition (always a winning coalition). Thus, the total number of winning coalitions is 2^{N-1}.

67. (a) In each nine-member winning coalition, every member is critical. In each coalition having 10 or more members, only the five permanent members are critical.

(b) At least nine members are needed to form a winning coalition. So, there are $210 + 638 = 848$ winning coalitions. Since every member is critical in each nine-member coalition, the nine-member coalitions yield a total of $210 \times 9 = 1890$ critical players. Since only the permanent members are critical in coalitions having 10 or more members, there are $638 \times 5 = 3190$ critical players in these coalitions. Thus, the total number of critical players in all winning coalitions is 5080.

(c) Each permanent member is critical in each of the 848 winning coalitions. Thus, the Banzhaf Power Index of a permanent member is 848/5080.

(d) The 5 permanent members together have $5 \times 848/5080 = 4240/5080$ of the power. The remaining 840/5080 of the power is shared equally among the 10 nonpermanent members, giving each a Banzhaf power index of 84/5080.

(e) In the given weighted voting system, the quota is 39, each permanent member has 7 votes, and each nonpermanent member has 1 vote. The total number of votes is 45 and so if any one of the permanent members does not vote for a measure there would be at most $45 - 7 = 38$ votes and the measure would not pass. Thus all permanent members have veto power. On the other hand, all 5 permanent members votes only add up to 35 and so at least 4 nonpermanent members votes are needed for a measure to pass.

Chapter 3

WALKING

A. Fair Division Concepts

1. **(a)** Let C = the value of the chocolate half (in Alex's eyes)
   S = the value of the strawberry half (in Alex's eyes)
   It is known that $C = 3S$ and $C + S = \$12$. Substituting,
   $$3S + S = \$12$$
   $$4S = \$12$$
   $$S = \$3.00$$
   $$C = 3(\$3) = \$9.00$$
   So, the chocolate half is worth \$9.00.

   (b) By (a), the strawberry half is worth \$3.00.

   (c) $\dfrac{60°}{180°} \times (\$3) + \dfrac{40°}{180°} \times (\$9)$
   $= \dfrac{1}{3} \times (\$3) + \dfrac{2}{9} \times (\$9)$
   $= \$3.00$

3. **(a), (b), (c)** Let C = the value of the chocolate part (in Kala's eyes)
   S = the value of the strawberry part (in Kala's eyes)
   V = the value of the vanilla part (in Kala's eyes)
   It is known that $S = 2V$, $C = 3V$, and $C + S + V = \$12$. Substituting,
   $$3V + 2V + V = \$12$$
   $$6V = \$12$$
   $$V = \$2.00$$
   $$S = 2 \times (\$2) = \$4.00$$
   $$C = 3 \times (\$2) = \$6.00$$

   (d) piece 1: $\dfrac{60°}{120°} \times (\$2) = \$1.00$; piece 2: $\dfrac{30°}{120°} \times (\$2) + \dfrac{30°}{120°} \times (\$4) = \$0.50 + \$1.00 = \$1.50$;
   piece 3: $\dfrac{60°}{120°} \times (\$4) = \$2.00$; piece 4: $\dfrac{30°}{120°} \times (\$4) + \dfrac{30°}{120°} \times (\$6) = \$1.00 + \$1.50 = \$2.50$;
   piece 5: $\dfrac{60°}{120°} \times (\$6) = \$3.00$; piece 6: $\dfrac{30°}{120°} \times (\$6) + \dfrac{30°}{120°} \times (\$2) = \$1.50 + \$0.50 = \$2.00$

5. **(a)** $\dfrac{1}{3} \times (\$12.00) = \$4.00$

   Any slice worth at least \$4.00 is a fair share to Ana. Slices s_2 and s_3 are fair shares to Ana.

(b) $\frac{1}{3} \times (\$15.00) = \$5.00$

Any slice worth at least $5.00 is a fair share to Ben. Slice s_3 is a fair share to Ben.

(c) $\frac{1}{3} \times (\$13.50) = \$4.50$

Any slice worth at least $4.50 is a fair share to Cara. Slices s_1, s_2, and s_3 are fair shares to Cara.

(d) Ben must receive s_3; So, Ana must receive s_2 and then Cara is left with s_1.

7. Any parcel worth at least 25% is a fair share.

   (a) Fair shares to Adams: s_1, s_4

   (b) Fair shares to Benson: s_1, s_2

   (c) Fair shares to Cagle: s_1, s_3

   (d) Fair shares to Duncan: s_4

   (e) Duncan must receive s_4 since that is the only parcel he considers to be a fair share. This leaves Adams with share s_1. Benson then receives s_2 and Cagle receives s_3.

9. **(a)** Let x denote the value of slices s_2 and s_3 to Abe. Then, $2x + 6.50 = 15.00$ and so $x = \$4.25$. Since Abe values a fair share at $\$15.00 / 4 = \$3.75$, he considers slices s_2 and s_3 to be fair shares.

   (b) Betty considers each share to be fair.

   (c) Let x denote the value (as a percentage of the whole cake) of slices s_1, s_2 and s_4 to Cory. Then, $x + x + 3x + x = 1$. So $x = 0.20$. The only share worth 25% of the cake is slice s_3 which is worth 60% of the cake to him.

   (d) Let x denote the value of slices s_2, s_3 and s_4 to Dana. Then, s_1 has a value of $x + \$2.00$. Also, $(x + \$2.00) + x + x + x = \$18.00$ and so $x = \$4$. Since Dana values a fair share at $\$18.00 / 4 = \$4.50$, she only considers slice s_1 to be a fair share.

   (e) Abe: s_2; Betty: s_4; Cory: s_3; Dana: s_1
   To start, Cory must receive s_3 and Dana must receive s_1. This forces Abe to receive s_2 in the division. Betty receives the remaining slice s_4.

B.   The Divider–Chooser Method

11. (a) Jared would divide the sandwich in two differently sized pieces – one piece would be all of the vegetarian part and 1/3 of the meatball part. The other would consist of 2/3 of the meatball part of the sandwich.

(b) Karla would divide the sandwich into one piece consisting of half of the vegetarian part (1/4 of the entire size of the sandwich) and another piece consisting of half of the vegetarian part and the entire meatball part (3/4 of the entire size of the sandwich).

(c) If Jared is the divider then he will cut the sandwich as described in part (a). Karla will obviously choose the vegetarian part. To her, that part is worth $8.00 (the meatball part of the sandwich is worthless to her). To Jared, the meatball part that he ends up with is worth $4.00 (exactly half the value of the sandwich).

(d) If Karla is the divider then she will cut the sandwich halfway down the vegetarian part (see part (b)). Jared will obviously choose the larger part containing the entire meatball part (3/4 of the size of the sandwich) for himself. To him, that share is worth $7.00 (the meatball half is worth $6.00 and he also gets half of the $2.00 vegetarian part). To Karla, the vegetarian part that she ends up with is worth $4.00 (exactly half the value of the sandwich).

13. (a) According to David, **(iii)** is the only one that shows a division of the pizza into fair shares because it is the only one in which the total amount of pepperoni, sausage, and mushroom is the same in each piece.

(b) Paul could choose either of the pieces in **(iii)** because the total amount of anchovies, mushroom, and pepperoni is the same in each piece.

15. (a) Answers may vary. For example, any of the following cuts are consistent with Paul's value system.

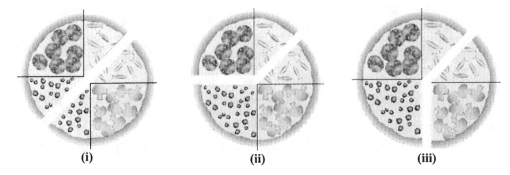

(i)   (ii)   (iii)

(b) David will choose the piece with the greatest total amount of pepperoni, sausage, and mushroom.
(i) either piece; **(ii)** II; **(iii)** I

17. (a) Since the two pieces are exact opposites, Mo must value pineapple and orange equally. Thus, the pineapple half is 50%.

(b) Let P = the value of the pineapple half (in Jamie's eyes)
O = the value of the orange half (in Jamie's eyes)

It is known that $O = 4P$ and $O + P = 100\%$.

$$4P + P = 100\%$$
$$5P = 100\%$$
$$P = 20\%$$
$$O = 4 \times 20\% = 80\%$$

left piece: $\dfrac{60°}{180°} \times 80\% + \dfrac{120°}{180°} \times 20\% = \dfrac{1}{3} \times 80\% + \dfrac{2}{3} \times 20\% = 40\%$

right piece: $\dfrac{120°}{180°} \times 80\% + \dfrac{60°}{180°} \times 20\% = \dfrac{2}{3} \times 80\% + \dfrac{1}{3} \times 20\% = 60\%$

(c) Jamie takes the right piece; Mo gets the left piece.

C.   The Lone-Divider Method

19. (a) If Divine is to receive s_1, then Chandra must receive s_3. So, Chase's fair share is s_2.

    (b) If Divine is to receive s_2, then Chase must receive s_3. So, Chandra's fair share is s_1.

    (c) If Divine is to receive s_3, then Chase must receive s_2 and Chandra must receive s_1.

21. (a) If DiPalma is to receive s_1, then Chou must receive s_4. So, Choate must receive s_3. This leaves Childs with s_2.

    (b) If DiPalma is to receive s_2, then Childs must receive s_3. So, Choate must receive s_4. This leaves Chou with s_1.

    (c) If DiPalma is to receive s_3, then Choate must receive s_4. So, Chou must receive s_1. This leaves Childs with s_2.

    (d) If DiPalma is to receive s_4, then Choate must receive s_3. So, Childs must receive s_2. This leaves Chou with s_1.

23. (a) If Cher is to receive s_2, then Chong must receive s_1. So, Cheech must receive s_3. This leaves Desi with s_4.

    (b) If Chong is to receive s_2, then Cher must receive s_3. So, Cheech must receive s_1. This leaves Desi with s_4.

    (c) None of the choosers chose s_4, which can only be given to the divider.

25. (a) If C_1 is to receive s_2, then C_2 must receive s_4. So, C_3 must receive s_3. It follows that C_4 must receive s_5. This leaves D with s_1.

(b) If C_1 is to receive s_4, then C_2 must receive s_2. So, C_3 must receive s_3. It follows that C_4 must receive s_5. This leaves D with s_1.

(c) There are only two possible cases: C_1 receives s_2 or C_1 receives s_4 (since those were the slices in C_1's bid list). The outcome of both cases appear in (a) and (b).

27. (a) A fair division of the cake is

C_1	C_2	C_3	C_4	C_5	D
s_5	s_1	s_6	s_2	s_3	s_4

(b) C_5 must get s_3 which forces C_4 to get s_2. This leaves only s_5 for C_1 which in turn leaves only s_6 for C_3. Consequently only s_1 is left for C_2 and the divider D must get s_4.

29. (a) The Divider was Gong since that is the only player that could possibly value each piece equally.

(b) To determine the bids placed, each player's bids should add up to $480,000. So, for example, Egan would bid $480,000 - $80,000 - $85,000 - $195,000 = $120,000 on parcel s_3. Further, one player must bid the same on each parcel (the divider). The following table shows the value of the four parcels in the eyes of each partner.

	s_1	s_2	s_3	s_4
Egan	$80,000	$85,000	$120,000	$195,000
Fine	$125,000	$100,000	$135,000	$120,000
Gong	$120,000	$120,000	$120,000	$120,000
Hart	$95,000	$100,000	$175,000	$110,000

For a player to bid on a parcel, it would need to be worth at least $480,000/4 = $120,000. Based on this table, the choosers would make the following declarations:
Egan: $\{s_3, s_4\}$; Fine: $\{s_1, s_3, s_4\}$; Hart: $\{s_3\}$.

(c) One possible fair division of the land is

Egan	**Fine**	**Gong**	**Hart**
s_4	s_1	s_2	s_3

(d) This is the only possible division of the land since Hart must receive s_3, so that Egan in turn must receive s_4. It follows that Fine must receive s_1 so that Gong (the divider) winds up with s_2.

D.   The Lone-Chooser Method

31. (a) Angela sees the left half of the cake as being worth $18. In her second division, she will create three $6 pieces.

$$\frac{x}{90°} \times \$13.50 = \$6.00$$

$$x = 40°$$

So Angela cuts two 40° pieces from the strawberry, and the remaining strawberry plus the vanilla makes up the third piece.

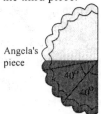

Angela's piece

(b) Boris sees the right half of the cake as being worth $15. In his second division, he will create three $5 pieces.

$$\frac{x}{90°} \times \$9.00 = \$5.00$$

$$x = 50°$$

So Boris cuts one 50° piece from the strawberry part of the cake. Also,

$$\frac{x}{90°} \times \$6.00 = \$5.00$$

$$x = 75°$$

So, Boris cuts one 75° piece from the vanilla part of the cake. The remaining 15° piece of vanilla plus the 40° piece of strawberry makes up the third piece.

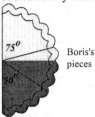

Boris's pieces

(c) Since Carlos values vanilla twice as much as strawberry, Carlos would clearly take the vanilla part of Angela's division. He would also want to take the most vanilla that he could from Boris. One possible fair division is

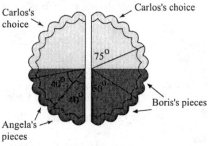

Carlos's choice

Carlos's choice

Boris's pieces

Angela's pieces

(d) Since she receives two of her $6.00 pieces, the value of Angela's final share (in Angela's eyes) is $12.00. Similarly, the value of Boris' final share (in Boris' eyes) is $10.00. The value of Carlos' final share (in Carlos' eyes) is

$$\frac{90°}{90°} \times \$12 + \frac{10°}{90°} \times \$6 + \frac{75°}{90°} \times \$12$$

$$= \$22.67.$$

33. Notice that the value of the strawberry piece (in Angela's eyes) is $\frac{120°}{180°} \times \$9 = \$6.00$.

The value of the other piece is, of course, also $6.00. Boris chooses the piece with all of the vanilla because he views it as being worth $12.00 + \$6.00 = \$18.00$.

(a) One possible second division by Boris is

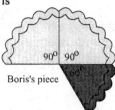

Boris's piece

Boris views his portion of the cake as

being worth $\$12 + \dfrac{60°}{180°} \times \$18 \approx \$18.00.$

So he will divide the cake into 3 pieces each worth $6.00.

(b) The second division by Angela is

Angela's piece

She divides the piece evenly.

(c) Carlos will select any one of Angela's pieces since they are identical in value to him. He will also select one of Boris's vanilla wedges since he values vanilla twice as much as he values strawberry. One possible fair division is

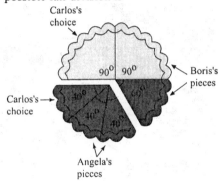

(d) Angela thinks her share is worth

$\dfrac{80°}{180°} \times \$27 = \$12.00$.

Boris thinks his share is worth

$\dfrac{90°}{180°} \times \$12 + \dfrac{60°}{180°} \times \$18 = \$12.00$.

Carlos thinks his share is worth

$\dfrac{40°}{180°} \times \$12 + \dfrac{90°}{180°} \times \$24 = \$14.67$.

35. After Arthur makes the first cut, Brian chooses the piece with the chocolate and strawberry, worth (to him) 100% of the cake.

(a) One possible second division by Brian is

Brian's piece

Brian likes chocolate and strawberry equally well, so he divides the piece evenly.

(b) The second division by Arthur is

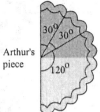

Arthur's piece

Arthur places all of the value on the orange half of the piece, so he divides the orange evenly.

(c) Since Carl likes chocolate and vanilla, one possible fair division is

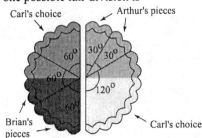

(d) Arthur thinks his share is worth $33\dfrac{1}{3}$ %.

Brian thinks his share is worth $66\dfrac{2}{3}$ %.

Carl thinks his share is worth

$\dfrac{60°}{90°} \times 50\% + 50\% = 83\dfrac{1}{3}\%$.

37. Suppose that Carl chooses the top half, worth (to him) 50% of the cake. (Carl could also have chosen the bottom half.)

(a) A second division by Carl is

Carl's piece

Carl places all of the value on the chocolate half of the piece, so he divides the chocolate evenly.

(b) A second division by Brian is

Brian's piece

Brian places all of the value on the strawberry half of the piece, so he divides the strawberry evenly.

(c) One possible fair division is

Arthur's choice

Carl's piece's

Brian's piece

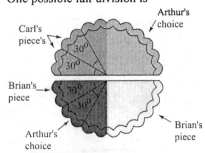

Arthur's choice

Brian's piece

(d) Arthur thinks his share is worth

$$50\% + \frac{30°}{90°} \times 50\% = 66\frac{2}{3}\%.$$

Brian thinks his share is worth $33\frac{1}{3}\%$.

Carl thinks his share is worth $33\frac{1}{3}\%$.

39. (a) Karla would divide the sandwich into one piece consisting of half of the vegetarian part (1/4 of the entire size of the sandwich) and another piece consisting of half of the vegetarian part and the entire meatball part (3/4 of the entire size of the sandwich). Jared would choose the larger slice containing the meatball part and half of the vegetarian part.

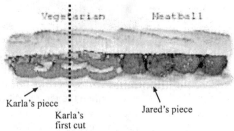

Karla's piece

Karla's first cut

Jared's piece

(b) Jared would divide his piece in three equally sized pieces – two pieces would be all meat (each 1/4 of the total sandwich) and the other would be all vegetarian (1/4 of the total sandwich).

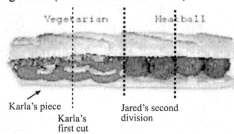

Karla's piece

Karla's first cut

Jared's second division

(c) Karla would divide her piece in three equally sized pieces – each piece would be all vegetarian (each 1/3 of 1/4 of the total sandwich).

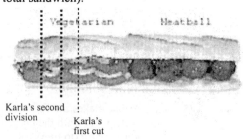

Karla's second division

Karla's first cut

(d) Since Lori, the chooser, likes the meatball part twice as much as the vegetarian part, she will select one of Jared's meatball parts and one of Karla's vegetarian parts.

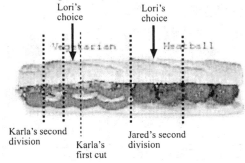

Jared will receive 1/2 of the total sandwich (part meatball and part vegetarian) which, since he views each part equally, he values at 1/2 of the total sandwich.

Karla will receive 2/3 of 1/2 (that is, 1/3) of the vegetarian part. She values this as worth 1/3 of the sandwich.

Lori will receive 1/2 of the meatball part and 1/3 of 1/2 (that is 1/6) of the vegetarian part. Since she views the meatball half of the sandwich as being worth 2/3 of the entire sandwich, she values her piece as $\dfrac{1}{2} \times \dfrac{2}{3} + \dfrac{1}{6} \times \dfrac{1}{3} = \dfrac{7}{18}$ of the entire sandwich.

E. The Last-Diminisher Method

41. (a) P_4; P_2 and P_4 are both diminishers since they value the piece when it is their turn as worth more than \$6.00. However, P_4 is the last diminisher.

(b) \$6.00; While P_4 values the piece as being worth \$6.50 when it is their turn to play, they diminish its value to \$6.00 when they make a claim.

(c) P_1; The remaining players are everyone except for P_4. They divide what remains into quarters.

43. (a) P_5; P_4 and P_5 are both diminishers since they value the piece when it is their turn as worth more than \$6.00. P_5 is the last diminisher.

(b) \$7.00; Since P_5 is the last player to play, they diminish the piece by 0%.

(c) P_1

45. (a) P_9 (the last diminisher)

(b) P_1

(c) P_5 (the last diminisher)

(d) P_1; There are no diminishers.

(e) P_2; P_1 already has a piece.

47. (a) Arthur's claim is a 45° wedge. He views the chocolate 90° wedge of the cake as being worth ½ of the total value of the cake. So ½ of that is worth ¼ of the value of the entire cake.

Arthur's claim

(b) No players are diminishers in round 1. Brian and Carl both view Arthur's claim as being worth exactly 1/4 of the cake. Damian only views Arthur's claim as being worth 1/8 of the cake.

(c) Since there were no diminishers, Arthur receives his original claim at the end of round 1.

(d) Brian could stake two different types of claims at the beginning of round 2. He could either claim the remaining 45° wedge of chocolate or he could claim a 45° wedge of strawberry (he views both as worth 1/3 of what remains of the cake). Neither Carl nor Damian will view such a piece as being worth more than 1/3 of the remaining cake (at best, Carl views a 45° wedge of chocolate as being worth 1/3 of what remains). So, Brian will receive his share at the end of round 2.

(e) There are two possibilities for what remains of the cake after round 2. Case I: All of the chocolate has been removed. In this case, Carl makes a claim on a 45° wedge of vanilla. Damian is happy to pass. So Carl will receive his share at the end of round 3. Case II: A 45° wedge of chocolate and a 45° wedge of strawberry have been removed. In this case, Carl makes a claim on a 67.5° wedge of vanilla. Damian is happy to pass and Carl will again receive his share at the end of round 3.

49. (a) Lori values the vegetarian part three times as much as she values the meatball part. Assume, for example, that she values the vegetarian part as being worth $3 and the meatball part as being worth $1. Then, she will cut a share worth 1/3 of the entire $4 sub. She will claim some fraction x of the $3 vegetarian part of the sandwich. Thus,

$$x(\$3) = \frac{1}{3}(\$4) \text{ making } x = \frac{4}{9}. \text{ So, 4/9}$$

of the vegetarian half of the sandwich will make up the *C*-piece.

(b) Karla will diminish the *C*-piece cut by Lori from 4/9 to 1/3 of the vegetarian part of the sandwich. Jared will pass and Karla is the last diminisher in round 1.

(c) Karla will receive 1/3 of the vegetarian part of the sandwich.

(d) At the end of round 1, 2/3 of the vegetarian part and the entire meatball part are left. Lori would value the vegetarian part at $2 and the meatball part at $1. She will claim some fraction x of the $2 vegetarian part of the remaining $3 sandwich. Thus,

$$x(\$2) = \frac{1}{2}(\$3) \text{ so that } x = \frac{3}{4}. \text{ So, 3/4}$$

of 2/3 of the vegetarian part of the sandwich will make up her *C*-piece. Jared will pass. Lori will receive a 1/2 of the vegetarian part of the sandwich (that is 1/4 of the entire sandwich).

(e) Jared will receive the entire meatball part of the sandwich and part of the vegetarian part of the sandwich. Since Karla receives 1/3 of the vegetarian part and Lori receives 1/2 of the vegetarian part, Jared will receive 1-1/3-1/2 = 1/6 of the vegetarian part in addition to the entire meatball part of the sandwich.

F.   The Method of Sealed Bids

51. (a) In the first settlement, Ana gets the desk and receives $120 in cash; Belle gets the; Chloe gets the vanity and the tapestry and pays $360.

Item	Ana	Belle	Chloe	
Dresser	150	**300**	275	
Desk	**180**	150	165	
Vanity	170	200	**260**	
Tapestry	400	250	**500**	
Total Bids	900	900	1200	
Fair share	300	300	400	Total
Value of items received	180	300	760	Surplus
Prelim cash settlement	120	0	−360	240

(b) Ana, Belle, and Chloe split the $240 surplus cash three ways. This adds $80 cash to the first settlement. In the final settlement, Ana gets the desk and receives $200 in cash; Belle gets the dresser and receives $80; Chloe gets the vanity and the tapestry and pays $280.

Item	Ana	Belle	Chloe	Surplus
Prelim cash settlement	120	0	−360	240
Share of surplus	80	80	80	
Final cash settlement	200	80	−280	

53. Bob gets the business and pays $155,000. Jane gets $80,000 and Ann gets $75,000.

Item	Bob	Ann	Jane	
Partnership	240000	210000	225000	
Total Bids	240000	210000	225000	
Fair Share	80000	70000	75000	Total
Value of items received	240000	0	0	Surplus
Prelim. cash settlement	−160000	70000	75000	15000
Share of surplus	5000	5000	50000	
Final cash settlement	−155000	75000	80000	

55. (a) There is a surplus of $130 after the first settlement in which *A* ends up with items 4 and 5, *C* ends up with items 1 and 3, and *E* ends up with items 2 and 6.

Item	*A*	*B*	*C*	*D*	*E*	
Item 1	352	295	**395**	368	324	
Item 2	98	102	98	95	**105**	
Item 3	460	449	**510**	501	476	
Item 4	**852**	825	832	817	843	
Item 5	**513**	501	505	505	491	
Item 6	725	738	750	744	**761**	
Total Bids	3000	2910	3090	3030	3000	
Fair Share	600	582	618	606	600	Total
Value of items rec'd	1365	0	905	0	866	Surplus
Prelim. cash	−765	582	−287	606	−266	130

(b) *A*, *B*, *C*, *D*, and *E* split the $130 surplus 5 ways ($26 each). *A* ends up with items 4 and 5 and pays $739; *B* ends up with $608; *C* ends up with items 1 and 3 and pays $261; *D* ends up with $632; *E* ends up with items 2 and 6 and pays $240.

57. Suppose that Angelina bid $*x* on the laptop. Then, she values the entire estate at *x*+$2900. Brad, on the other hand, values the entire estate at $4640. The rest of the story is given in the table below.

	Angelina	Brad
Fair Share	(*x*+$2900)/2	$2320
Value of items rec'd	*x*+$300	$2780
Prelim. cash	(*x*+$2900)/2-(*x*+$300)	-$460
Share of surplus	$105	$105
Final cash	$355	-$355

To determine the value of *x*, it must be that $\dfrac{\$2300-x}{2}+\$105=\$355$. This leads to $\$2300-x=\$500$ and *x* = $1800. Angelina bid $1800 on the laptop.

G.   The Method of Markers

59. (a) First *B* gets items 1, 2, 3; Then, *C* gets items 5, 6, 7; Finally, *A* gets items 10, 11, 12, 13.

(b) Items 4, 8, and 9 are left over.

61. (a) First *A* gets items 1, 2; Then, *C* gets items 4, 5, 6, 7; Finally, *B* gets items 10, 11, 12.

(b) Items 3, 8, and 9 are left over.

63. (a) First C gets items 1, 2, 3; Then, E gets items 5, 6, 7, 8; Then, D gets items 11, 12, 13; Then, B gets items 15, 16, 17; Finally, A gets items 19, 20.

   (b) Items 4, 9, 10, 14, and 18 are left over.

65. (a) Quintin thinks the total value is $3 \times \$12 + 6 \times \$7 + 6 \times \$4 + 3 \times \$6 = \$120$, so to him a fair share is worth \$30. Ramon thinks the total value is $3 \times \$9 + 6 \times \$5 + 6 \times \$5 + 3 \times \$11 = \$120$, so to him a fair share is worth \$30. Stephone thinks the total value is $3 \times \$8 + 6 \times \$7 + 6 \times \$6 + 3 \times \$14 = \$144$, so to him a fair share is worth \$36. Tim thinks the total value is $3 \times \$5 + 6 \times \$4 + 6 \times \$4 + 3 \times \$7 = \$84$, so to him a fair share is worth \$21. They would place their markers as shown below.

   (b)

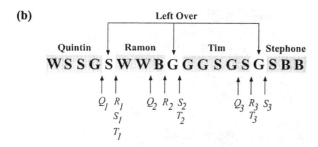

   (c) A coin toss would be needed to decide if Quentin or Stephone should receive S. If Stephone were to receive S, then Stephone would receive S and both G's and would pay \$13.32. In that case, Quentin and Ramon would receive \$4.69 and Tim would receive \$3.94.

67. (a) Ana places her markers between the different types of candy bars. Belle places her markers between each of the Nestle Crunch bars. Chloe places her markers thinking that each Snickers and Nestle Crunch bar is worth \$1 and each Reese's is worth \$2. Since Chloe would then value the bars to be worth \$12 in total, she would place markers after each \$4 worth of bars.

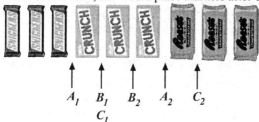

(b) Two Nestle Crunch bars and one Reese's Peanut Butter Cup is left over.

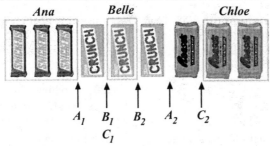

(c) Belle would select one of the Nestle Crunch bars, Chloe would then select the Reese's Peanut Butter Cup, and then Ana would be left with the last Nestle Crunch bar.

JOGGING

69. (a) Two divisions are possible.

Abe	Babe	Cassie
P, S	*Q*	*R*
P, S	*R*	*Q*

(b) Four divisions are possible.

Abe	Babe	Cassie
R, S	*Q*	*P*
P, R	*Q*	*S*
Q, S	*R*	*P*
P, Q	*R*	*S*

(c) Out of P, Q, and S, Abe gets two of the three. Combine the left over one with R. Since Babe and Cassie each consider Abe's pieces to be worth less than $1.50 each, they much think the rest is worth more than $3.00, and thus, they can divide it fairly using the divider-chooser method.

71. (a) The total area is $30,000\ m^2$ and the area of C is only $8000\ m^2$. Since P_2 and P_3 value the land uniformly, each thinks that a fair share must have an area of at least $10,000\ m^2$.

(b) Since there are 22,000 m^2 left, any cut that divides the remaining property in parts of 11,000 m^2 will work. For example,

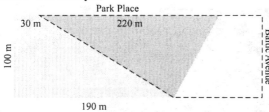

(c) The cut parallel to Park Place which divides the parcel in half is illustrated below. The cut is made x meters from the bottom. We know that $\frac{y}{x} = \frac{60}{100}$ so that $y = \frac{3}{5}x$. The bottom trapezoid is to have area of 11,000 m^2. So, $\frac{(190+(3/5)x)+190}{2} \times x = 11,000$ or $3x^2 + 1900x - 110,000 = 0$. By the quadratic formula, the value of x is $\frac{-950+50\sqrt{493}}{3} \approx 53.4$ m.

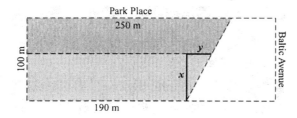

(d) The cut parallel to Baltic Avenue which divides the parcel in half is

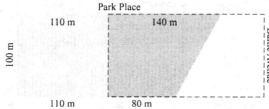

73. (a) If P_2 and P_3 each value the land (except for the 20m by 20m plot in the upper left corner) at v dollars per square meter, then the total value of the land to each of them is $(300)(100)v - 2(20)(20)v = 29,200v$. Thus a fair share to each of them would have value at least 9,733.33v. But the value of piece C to each of them is $100\left(\frac{70+90}{2}\right)v = 8000v$ and so they would both pass and P_1 would end up with piece C.

(b) Since the value of the remaining land to each of P_2 and P_3 is $29{,}200v - 8000v = 21{,}200v$, the

divider in round 2 wants to make the cut so that each piece is worth $\dfrac{21{,}200}{2}v = 10{,}600v$. Letting x

denote the length of the shorter base of the trapezoid II as shown in the figure,

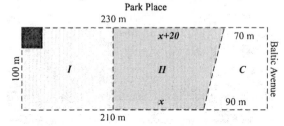

the value of piece II to each of them is $100\left(\dfrac{x+(x+20)}{2}\right)v = 100(x+10)v$ and consequently x must

satisfy the equation $100(x+10)v = 10{,}600v$ which gives $x = 96$.

75. (a) No. The chooser is not assured a fair share in her own value system.

(b) The divider. The divider has the advantage that he can assure himself a fair share of the cake if he divides it appropriately. The chooser is essentially at the mercy of the laws of chance.

77. (a) *A*-Allen, Bryant; *B*-Evans, Francis; *C*-Carter, Duncan
First pick: *A*-Allen; Second pick: *B*-Evans; Third pick: *C*-Carter; Fourth pick: *A*-Bryant; Fifth pick: *B*-Francis; Sixth pick: *C*-Duncan

(b) *A*-Allen, Duncan; *B*-Evans, Bryant; *C*-Carter, Francis
First pick: *A*-Allen; Second pick: *B*-Evans; Third pick: *C*-Carter; Fourth pick: *C*-Francis; Fifth pick: *B*-Bryant; Sixth pick: *A*-Duncan

(c) No. In (a), *A* and *B* do as well as they can. In (b), *C* does as well as they can and cannot be improved. This means that *B* cannot be improved since Francis must go to *C*.

Chapter 4

WALKING

A.   Standard Divisors and Quotas

1. (a) Standard divisor $= \dfrac{3,310,000 + 2,670,000 + 1,330,000 + 690,000}{160} = 50,000$

(b) Apure: $\dfrac{3,310,000}{50,000} = 66.2$;

Barinas: $\dfrac{2,670,000}{50,000} = 53.4$;

Carabobo: $\dfrac{1,330,000}{50,000} = 26.6$;

Dolores: $\dfrac{690,000}{50,000} = 13.8$

(c)

State	Apure	Barinas	Carabobo	Dolores
Upper Quota	67	54	27	14
Lower Quota	66	53	26	13

3. (a) The "states" in any apportionment problem are the entities that will have "seats" assigned to them according to a share rule. In this case, the states are the 6 bus routes and the seats are the 130 buses.

(b) Standard divisor $= \dfrac{45,300 + 31,070 + 20,490 + 14,160 + 10,260 + 8,720}{130} = 1000$

The standard divisor represents the average number of passengers per bus per day.

(c) A: $\dfrac{45,300}{1000} = 45.30$;

B: $\dfrac{31,070}{1000} = 31.07$;

C: $\dfrac{20,490}{1000} = 20.49$;

D: $\dfrac{14,160}{1000} = 14.16$;

E: $\dfrac{10,260}{1000} = 10.26$;

F: $\dfrac{8,720}{1000} = 8.72$

5. (a) Number of seats $= 40.50 + 29.70 + 23.65 + 14.60 + 10.55 = 119$

    (b) Standard divisor $= \dfrac{23,800,000}{119} = 200,000$

    (c) Population = standard quota × standard divisor, so
        A: $40.5 \times 200,000 = 8,100,000$
        B: $29.7 \times 200,000 = 5,940,000$
        C: $23.65 \times 200,000 = 4,730,000$
        D: $14.60 \times 200,000 = 2,920,000$
        E: $10.55 \times 200,000 = 2,110,000$

7. Standard quota of Texas = 32.32.
    With 7.43% of the U.S. population, Texas should receive 7.43% of the number of seats available in the House of Representatives. So, the standard quota for Texas is $0.0743 \times 435 = 32.3205$.

9. (a) Standard divisor $= \dfrac{100\%}{200} = 0.5\%$

    (b)

State	A	B	C	D	E	F
Standard Quota	22.74	16.14	77.24	29.96	20.84	33.08

B.   Hamilton's Method

11. Lower Quotas are:
    A : 66
    B : 53
    C : 26
    <u>*D* : 13</u>
    Sum is 158. So we have 160–158 = 2 seats remaining to allocate. These are given to *D* and *C*, since they have the largest fractional parts of the standard quota. The final apportionment is:
    A: 66; *B*: 53; *C*: 27; *D*: 14.

13. Lower Quotas are:
    A: 45; *B*: 31; *C*: 20; *D*: 14; *E*: 10; *F*: 8, and the sum is 128. So we have 130–128 = 2 buses remaining to allocate. These are given to *F* and *C*, since they have the largest fractional parts of the standard quota. The final apportionment is:
    A: 45; *B*: 31; *C*: 21; *D*: 14; *E*: 10; *F*: 9.

15. Lower Quotas are:
    A: 40; *B*: 29; *C*: 23; *D*: 14; *E*: 10, and the sum is 116. So we have 119–116 = 3 seats remaining to allocate. These are given to *B*, *C* and *D*, since they have the largest fractional parts of the standard quota. The final apportionment is:
    A: 40; *B*: 30; *C*: 24; *D*: 15; *E*: 10.

17. Lower Quotas are:
    A: 22; *B*: 16; *C*: 77; *D*: 29; *E*: 20, *F*: 33, and the sum is 197. So we have 200–197 = 3 seats remaining to allocate. These are given to *A*, *D* and *E*, since they have the largest fractional parts of the standard quota. The final apportionment is:
    A: 23; *B*: 16; *C*: 77; *D*: 30; *E*: 21; *F*: 33.

19. **(a)**

Child	Bob	Peter	Ron
Standard quota	0.594	2.673	7.733
Lower quota	0	2	7

Note: standard divisor: $= \dfrac{54 + 243 + 703}{11} = 90.\overline{90}$

The sum of lower quotas is 9, so there are 2 remaining pieces of candy to allocate. These are given to Ron and Peter, since they have the largest fractional parts of the standard quota.
The final apportionment is:
Bob: 0; Peter: 3; Ron: 8.

(b)

Child	Bob	Peter	Ron
Study time	56	255	789
Standard quota	.56	2.55	7.89
Lower quota	0	2	7

Note: standard divisor $= \dfrac{56 + 255 + 789}{11} = 100$

The sum of the lower quotas is 9, so there are 2 remaining pieces of candy to allocate. These are given to Ron and Bob, since they have the largest fractional parts of the standard quota.
The final apportionment is:
Bob: 1; Peter: 2; Ron: 8.

(c) Yes. For studying an extra 2 minutes (an increase of 3.70%), Bob gets a piece of candy. However, Peter, who studies an extra 12 minutes (an increase of 4.94%), has to give up a piece. This is an example of the population paradox.

21. **(a)**

Child	Bob	Peter	Ron
Standard quota	.594	2.673	7.733
Lower quota	0	2	7

Note: standard divisor: $= \dfrac{54 + 243 + 703}{11} = 90.\overline{90}$

The sum of lower quotas is 9, so there are 2 remaining pieces of candy to allocate. These are given to Ron and Peter, since they have the largest fractional parts of the standard quota.
The final apportionment is:
Bob: 0; Peter: 3; Ron: 8.

(b)

Child	Bob	Peter	Ron	Jim
Study time	54	243	703	580
Standard quota	0.58	2.61	7.56	6.24
Lower quota	0	2	7	6

Note: standard divisor $= \dfrac{54 + 243 + 703 + 580}{17} \approx 92.94$

The sum of the lower quotas is 15, so there are 2 remaining pieces of candy to allocate. These are given to Peter and Bob, since they have the largest fractional parts of the standard quota.
The final apportionment is:
Bob: 1; Peter: 3; Ron: 7; Jim: 6.

(c) Ron loses a piece of candy to Bob when Jim enters the discussion and is given his fair share (6 pieces) of candy. This is an example of the new-states paradox.

C.   Jefferson's Method

23. Any modified divisor between approximately 49,285.72 and 49,402.98 can be used for this problem. Using $D = 49{,}300$ we obtain:

State	A	B	C	D
Modified Quota	67.14	54.16	26.98	13.996
Modified Lower Quota	67	54	26	13

The final apportionment is the modified lower quota value in the table.

25. Any modified divisor between approximately 971 and 975.7 can be used for this problem. Using $D = 975$ we obtain:

Route	A	B	C	D	E	F
Modified Quota	46.46	31.87	21.02	14.52	10.52	8.94
Modified Lower Quota	46	31	21	14	10	8

The final apportionment is the modified lower quota value in the table.

27. Any modified divisor between approximately 194,666.67 and 197,083.33 can be used for this problem. Using $D = 195{,}000$ we obtain:

State	A	B	C	D	E
Modified Quota	41.54	30.46	24.26	14.97	10.82
Modified Lower Quota	41	30	24	14	10

The final apportionment is the modified lower quota value in the table.

29. Any modified divisor between approximately 0.4944% and 0.4951 can be used for this problem. Using $D = 0.495\%$ we obtain:

State	A	B	C	D	E	F
Modified Quota	22.9697	16.3030	78.0202	30.2626	21.0505	33.4141
Modified Lower Quota	22	16	78	30	21	33

The final apportionment is the modified lower quota value in the table.

31. From Exercise 7, the standard quota of Texas was 32.32 for the 2000 Census. However, Jefferson's method only allows for violations of the upper quota. This means that Texas could not receive fewer than 32 seats under Jefferson's method.

D.   **Adams' Method**

33. Any modified divisor between approximately 50,377.4 and 50,923 can be used for this problem. Using $D = 50,500$ we obtain:

State	A	B	C	D
Modified Quota	65.54	52.87	26.34	13.66
Modified Upper Quota	66	53	27	14

The final apportionment is the modified upper quota value in the table.

35. Any modified divisor between approximately 1024.6 and 1026 can be used for this problem. Using $D = 1025$ we obtain:

Route	A	B	C	D	E	F
Modified Quota	44.20	30.31	19.99	13.81	10.01	8.51
Modified Upper Quota	45	31	20	14	11	9

The final apportionment is the modified upper quota value in the table.

37. Any modified divisor between approximately 204,828 and 205,652 can be used for this problem. Using $D = 205,000$ we obtain:

State	A	B	C	D	E
Modified Quota	39.51	28.98	23.07	14.24	10.29
Modified Upper Quota	40	29	24	15	11

The final apportionment is the modified upper quota value in the table.

39. Any modified divisor between approximately 0.5044% and 0.5081% can be used for this problem. Using $D = 0.505\%$ we obtain:

State	A	B	C	D	E	F
Modified Quota	22.5149	15.9802	76.4752	29.6633	20.6337	32.7525
Modified Upper Quota	23	16	77	30	21	33

The final apportionment is the modified upper quota value in the table.

41. The difference between 50 (the number of seats that California would receive under Adams' method) and 52.45 (California's standard quota) is greater than 1. This fact illustrates that Adams' method violates the quota rule.

E.   Webster's Method

43. Any modified divisor between approximately 49,907 and 50,188 can be used for this problem. Using $D = 50,000$ we obtain:

State	A	B	C	D
Modified Quota	66.2	53.4	26.6	13.8
Rounded Quota	66	53	27	14

The final apportionment is the rounded quota value in the table.

45. Any modified divisor between approximately 995.61 and 999.51 can be used for this problem. Using $D = 996$ we obtain:

Route	A	B	C	D	E	F
Modified Quota	45.48	31.19	20.57	14.22	10.30	8.76
Rounded Quota	45	31	21	14	10	9

The final apportionment is the rounded quota value in the table.

47. Any modified divisor between approximately 200,953 and 201,276 can be used for this problem. Using $D = 201,000$ we obtain:

State	A	B	C	D	E
Modified Quota	40.30	29.55	23.53	14.53	10.498
Rounded Quota	40	30	24	15	10

The final apportionment is the rounded quota value in the table.

49. Any modified divisor between approximately 0.499% and 0.504% can be used for this problem. Using $D = 0.5\%$ we obtain:

State	A	B	C	D	E	F
Modified Quota	22.74	16.14	77.24	29.96	20.84	33.08
Rounded Quota	23	16	77	30	21	33

The final apportionment is the rounded quota value in the table.

JOGGING

51. **(a)** $q_1 + q_2 + \ldots + q_N$ represents the total number of seats available.

   (b) $\dfrac{P_1 + P_2 + \ldots + P_N}{q_1 + q_2 + \ldots + q_N}$ represents the total population divided by the total number of seats available which also happens to be the standard divisor.

   (c) $\dfrac{P_N}{P_1 + P_2 + \ldots + P_N}$ represents the percentage of the total population in the Nth state.

53. **(a)** Take for example $q_1 = 3.9$ and $q_2 = 10.1$ (with $m = 14$). Under both Hamilton's method and Lowndes' method, A gets 4 seats and B gets 10 seats.

   (b) Take for example $q_1 = 3.4$ and $q_2 = 10.6$ (with $m = 14$). Under Hamilton's method, A gets 3 seats and B gets 11 seats. Under Lowndes' method, A gets 4 seats and B gets 10 seats.

   (c) Assume that $f_1 > f_2$, so under Hamilton's method the surplus seat goes to A. Under Lowndes' method, the surplus seat would go to B if
   $$\frac{f_2}{q_2 - f_2} > \frac{f_1}{q_1 - f_1}$$
   which can be simplified to
   $$f_2(q_1 - f_1) > f_1(q_2 - f_2)$$
   $$f_2 q_1 - f_2 f_1 > f_1 q_2 - f_1 f_2$$
   $$f_2 q_1 > f_1 q_2$$
   $$\frac{q_1}{q_2} > \frac{f_1}{f_2} \text{ since all values are } > 0.$$

55. (a) In Jefferson's method the modified quotas are larger than the standard quotas and so rounding downward will give each state at least the integer part of the standard quota for that state.

(b) In Adam's method the modified quotas are smaller than the standard quota and so rounding upward will give each state at most one more than the integer part of the standard quota for that state.

(c) If there are only two states, an upper quota violation for one state results in a lower quota violation for the other state (and vice versa). Since neither Jefferson's nor Adams' method can have both upper and lower violations of the quota rule, neither can violate the quota rule when there are only two states.

57. (a) The standard divisor is given by $\dfrac{1,262,505}{7.671} \approx 164,581.54$. This means each representative represents (roughly) 164,582 people. Since the number of seats is M=300, the U.S. population can be estimated as $300 \times 164,581.54 \approx 49,374,462$.

(b) Since the standard quota for Texas is defined as the population of Texas divided by the standard divisor, it follows that the population of Texas is the product of its standard quota (9.672) and the standard divisor (164,581.54). That is, the population of Texas is $9.672 \times 164,581.54 \approx 1,591,833$.

59. (a) Any modified divisor between approximately 93.8 and 95.2 can be used for this problem. Using $D = 95$ we obtain:

State	A	B	C	D
Modified Quota	5.26	10.53	15.79	21.05
Modified Lower Quota	5	10	15	21

The final apportionment is the modified lower quota value in the table.

(b) For $D = 100$, the modified quotas are A: 5, B: 10, C: 15, D: 20; which sums to 50 seats. For $D < 100$, each of the modified quotas will increase, so rounding upward will give at least A: 6, B: 11, C: 16, D: 21 for a total of at least 54 seats. But for $D > 100$, each of the modified quotas will decrease, and so rounding upward will give at most A: 5, B: 10, C: 15, D: 20 for a total of 50 at most.

(c) From part (b), we see that there is no divisor such that the sum of the modified upper quotas will be 51.

61. Apportionment #1: Webster's method; Apportionment #2: Adams' method; Apportionment #3: Jefferson's method.
Adams' method favors small states and Jefferson's method favors large states.

Mini-Excursion 1

A.   The Geometric Mean

1. **(a)** $G = \sqrt{10 \times 100} = \sqrt{1000} = 100$

  (b) $G = \sqrt{20 \times 2000} = \sqrt{40,000} = 200$

  (c) $G = \sqrt{\dfrac{1}{20} \times \dfrac{1}{2000}} = \sqrt{\dfrac{1}{40,000}} = \dfrac{1}{\sqrt{40,000}} = \dfrac{1}{200}$

3. $G = \sqrt{2 \cdot 3^4 \cdot 7^3 \cdot 11 \cdot 2^7 \cdot 7^5 \cdot 11} = \sqrt{2^8 \cdot 3^4 \cdot 7^8 \cdot 11^2} = 2^4 \cdot 3^2 \cdot 7^4 \cdot 11$

5. Approximately 18.6%

To compute the average annual increase in home prices, we first find the geometric mean of 1.117 and 1.259. This is given by $G = \sqrt{1.117 \times 1.259} = \sqrt{1.406303} \approx 1.186$. So, the average annual increase in home prices is approximately 18.6%.

7.

Consecutive integers	Geometric mean (G)	Arithmetic mean (A)	Difference (A-G)
15,16	15.492	15.5	0.008
16,17	16.492	16.5	0.008
17,18	17.493	17.5	0.007
18,19	18.493	18.5	0.007
19,20	19.494	19.5	0.006
29,30	29.496	29.5	0.004
39,40	39.497	39.5	0.003
49,50	49.497	49.5	0.003

9. **(a)** Since $G = \sqrt{a \cdot b}$, the geometric mean of $k \cdot a$ and $k \cdot b$ is $\sqrt{ka \cdot kb} = k\sqrt{a \cdot b} = kG$.

  (b) Since $G = \sqrt{a \cdot b}$, the geometric mean of $\dfrac{k}{a}$ and $\dfrac{k}{b}$ is $\sqrt{\dfrac{k}{a} \cdot \dfrac{k}{b}} = k\dfrac{1}{\sqrt{a \cdot b}} = \dfrac{k}{G}$.

11. (a) Since $b > a$ and

$$|AC|^2 - |AB|^2 = \left(\frac{b+a}{2}\right)^2 - \left(\frac{b-a}{2}\right)^2$$

$$= \frac{b^2 + 2ab + a^2 - b^2 + 2ab - a^2}{4}$$

$$= ab$$

the length of BC is given by $\sqrt{ab}$.

(b) Since the length of the legs of a right triangle are always strictly less than the length of the hypotenuse, it follows that $\sqrt{ab} < \dfrac{a+b}{2}$.

B.   **The Huntington-Hill Method**

13. In this example the Huntington-Hill method produces the same apportionment as Webster's method and in fact, the standard divisor can be used as an appropriate divisor.

State	A	B	C	D	E	Total
Std. Quota	25.26	18.32	2.58	37.16	40.68	124
Geo. Mean (cutoff)	$\sqrt{25 \times 26} \approx$ 25.4951	$\sqrt{18 \times 19} \approx$ 18.4932	$\sqrt{2 \times 3} \approx$ 2.4495	$\sqrt{37 \times 38} \approx$ 37.4967	$\sqrt{40 \times 41} \approx$ 40.4969	
Apportionment	25	18	3	37	41	124

15. (a) Under Webster's method the standard divisor $D = 10,000$ works!

State	A	B	C	D	E	F	Total
Population	344,970	408,700	219,200	587,210	154,920	285,000	2,000,000
Standard Quota	34.497	40.87	21.92	58.721	15.492	28.5	
Apportionment	34	41	22	59	15	29	200

(b) Under the Huntington-Hill method the standard divisor does not work! However, a modified divisor of $D = 10,001$ does work.

State	A	B	C	D	E	F	Total
Population	344,970	408,700	219,200	587,210	154,920	285,000	2,000,000
Mod. Quota (D = 10,001)	34.494	40.866	21.918	58.715	15.490	28.497	
Geo. Mean	34.496	40.497	21.494	58.498	15.492	28.496	
Apportionment	34	41	22	59	15	29	200

(c) The apportionments came out the same for both methods.

17. (a) Under the Huntington-Hill method the standard divisor does not work but a modified divisor of $D = 990$ works!

State	Aleta	Bonita	Corona	Doritos	Total
Population	86,915	4,325	5,400	3,360	100,000
Modified Quota (D = 990)	87.793	4.369	5.455	3.394	
Geo. Mean	87.4986	4.4721	5.4772	3.4641	
Apportionment	88	4	5	3	100

(b) The apportionment using the Huntington-Hill method violates the quota rule (an upper quota violation). Aleta has a standard quota of 86.915 and yet receives 88 representatives in the final apportionment.

Chapter 5

A.   Graphs: Basic Concepts

1. (a) Vertex set: V = {A, B, C, X, Y, Z};
Edge set: E = {AX, AY, AZ, BX, BY, BZ,
CX, CY, CZ};
deg(A) = 3, deg(B) = 3, deg(C) = 3,
deg(X) = 3, deg(Y) = 3, deg(Z) = 3.

(b) Vertex set: V = {A, B, C};
Edge set: E = { };
deg(A) = 0, deg(B) = 0, deg(C) = 0.

(c) Vertex set: V = {V, W, X, Y, Z};
Edge set: E = {XX, XY, XZ, XV, XW,
WY, YZ};
deg (V) = 1, deg (W) = 2, deg(X) = 6,
deg(Y) = 3, deg(Z) = 2.

3. (a)

$$A \quad B \quad C \quad D$$

(b)

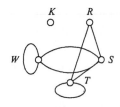

5. (a) Both graphs have the same vertex set
V = {A, B, C, D} and the same edge set
E = {AB, AC, AD, BD}.

(b)

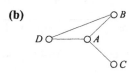

7. (a)

(b)

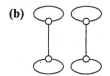

(c)

9. (a)

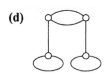

(b)

(c)

(d)

11. (a) C, B, A, H, F

(b) C, B, D, A, H, F

(c) C, B, A, H, F

(d) C, D, B, A, H, G, G, F

(e) 4 (*C, B, A*; *C, D, A*; *C, B, D A*; *C, D, B, A*)

(f) 3 (*H, F*; *H, G, F*; *H, G, G, F*)

(g) 12 (Any one of the paths in (e) followed by any one of the paths in (f).)

13. Circuits of length 1: *E, E*
Circuits of length 2: *B*, C, *B*
Circuits of length 3: *A, B, C, A;*
         A, B, C, A
[There are two edges between *B* and *C*.]
Circuits of length 4: *A, B, E, D, A*
Circuits of length 5: *A, B, E, E, D, A;*
         A, C, B, E, D, A;
         A, C, B, E, D, A;
Circuits of length 6: *A, C, B, E, E, D, A;*
         A, C, B, E, E, D, A;
         A, B, C, B, E, D, A
Circuits of length 7: *A, B, C, B, E, E, D, A*

B.   Graph Models

15.

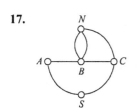

17.

19.

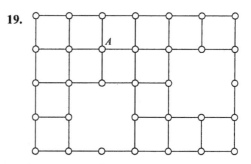

C.   Euler's Theorems

21. (a) Has an Euler circuit because all vertices have even degree.

(b) Has neither an Euler circuit nor an Euler path because there are four vertices of odd degree.

23. (a) Has neither an Euler circuit nor an Euler path because there are more than two (10 in fact) vertices of odd degree.

(b) Has no Euler circuit, but has an Euler path because there are exactly two vertices of odd degree.

(c) Has neither an Euler circuit nor an Euler path because the graph is not connected.

25. (a) Has neither an Euler circuit nor an Euler path because there are eight vertices of odd degree.

(b) Has no Euler circuit, but has an Euler path because there are exactly two vertices of odd degree.

(c) Has no Euler circuit, but has an Euler path because there are exactly two vertices of odd degree.

D.   Finding Euler Circuits and Euler Paths

27.

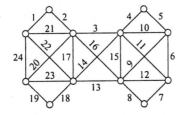

29. There are many possible Euler circuits. One possible circuit is given by:

$M, N, G, F, N, K, A, B, C, D, E, F, G, H, I,$
$J, A, J, K, O, D, E, O, L, H, I, L, M, C, B, M$

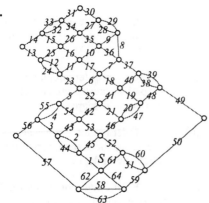

31.

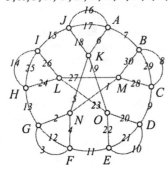

33.

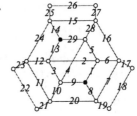

35. (a)

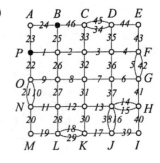

(b)

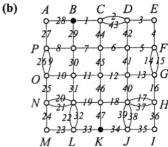

E.   Unicursal Tracings

37. (a) The drawing has neither because there are more than two vertices of odd degree.

116

(b) The drawing has an open unicursal tracing. For example,

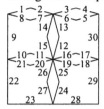

(c) The drawing has an open unicursal tracing. For example,

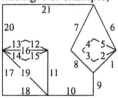

39. (a) The drawing has an open unicursal tracing. For example,

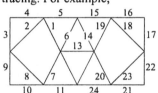

(b) The drawing has an open unicursal tracing. For example,

(c) The drawing has neither an open unicursal tracing nor a closed unicursal tracing because there are more than two vertices of odd degree.

F.   Eulerizations and Semi-Eulerizations

41. (a)

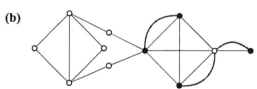

(b)

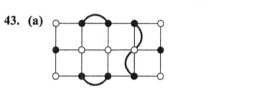

43. (a)

(b)

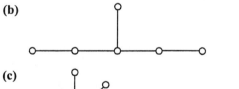

G. Miscellaneous

45. (a)

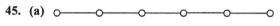

(b)

(c)

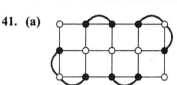

47. (a)

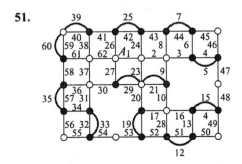

(b) Edges *BC* and *JK* will need to be retraced.

49. You would need to lift your pencil 4 times. There are 10 vertices of odd degree. Two can be used as the starting and ending vertices. The remaining eight odd degree vertices can be paired so that each pair forces one lifting of the pencil.

51.

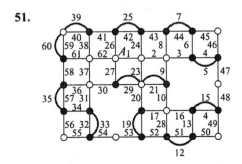

JOGGING

53. (a) An edge *XY* contributes 2 to the sum of the degrees of all vertices (1 to the degree of *X*, and 1 to the degree of *Y*).

(b) If there were an odd number of vertices, the sum of the degrees of all the vertices would be odd.

55. (a) If each vertex were of odd degree, then the graph has an odd number of odd vertices. This is not possible! So, it must be that each vertex is in fact of even degree. By Euler's Circuit Theorem, the graph would have an Euler circuit.

(b) The following graph is regular (all vertices are of degree 2) and it has an Euler circuit.

The graph below is also regular (all vertices have degree 3) but it does not have an Euler circuit.

57. Recall that in a connected graph, a bridge is an edge such that if we were to erase it, the graph would become disconnected. Also, remember that a connected graph having all even vertices must have an Euler circuit. But if a graph has an Euler circuit, it cannot have a bridge (because then it would be impossible to get from one component of the graph to another and back).

59. (a) 12 (See edges that have been added in part (b).)

(b)

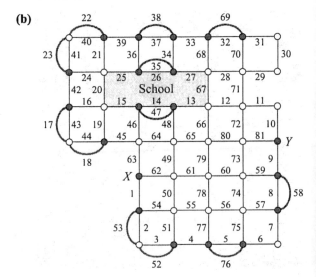

61. (a) The best graphs to build for $N = 4$, $N = 5$, and $N = 6$ vertices are illustrated below. The strategy that will give you the most money is to connect all but one of the vertices with a complete set of edges and to leave a single vertex isolated (incident to no edges).

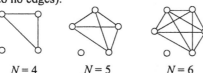

$N = 4$ $N = 5$ $N = 6$

(b) Using the strategy found in part (a), the most money you can make in building a graph with N vertices is
$$1 + 2 + 3 + \ldots + (N-3) + (N-2) =$$
$$\frac{(N-2)(N-1)}{2}.$$ [For the first vertex, there are N-2 edges that can be built, for the second vertex, there are only N-3 edges that can be built, etc…]

63. (a) Since vertices R, B, D, and L are all odd, the starting and ending vertices must be selected from this list. Moreover, a route will be optimal if only one bridge needs to be crossed twice. There are many ways to do that. For example, starting at R and ending at B and crossing the Adams Bridge twice produces an optimal route: R, L, R, A, C, L, C, D, L, D, B, R, B.
Listing the bridges in the order they are crossed: Washington, Jefferson, Truman, Wilson, Grant, Roosevelt, Monroe, Adams, Adams, Kennedy, Hoover, Lincoln.

(b) If the route must start at B and end at L, then one possible optimal route would cross the Lincoln Bridge twice and the Kennedy Bridge twice: B, R, B, R, L, R, A, C, L, C, D, B, D, L.
Listing the bridges in the order they are crossed: Hoover, Lincoln, Lincoln, Washington, Jefferson, Truman, Wilson, Grant, Roosevelt, Monroe, Kennedy, Kennedy, Adams.

Chapter 6

WALKING

A. Hamilton Circuits and Hamilton Paths

1. (a) 1. *A, B, D, C, E, F, G, A*;
     2. *A, D, C, E, B, G, F, A*;
     3. *A, D, B, E, C, F, G, A*

(b) *A, G, F, E, C, D, B*

(c) *D, A, G, B, C, E, F*

3. 1. *A, B, C, D, E, F, G, A*
   2. *A, B, E, D, C, F, G, A*
   3. *A, F, C, D, E, B, G, A*
   4. *A, F, E, D, C, B, G, A*

Mirror-image circuits:
5. *A, G, F, E, D, C, B, A*
6. *A, G, F, C, D, E, B, A*
7. *A, G, B, E, D, C, F, A*
8. *A, G, B, C, D, E, F, A*

The reasons that the above are the only Hamilton circuits are as follows. First note that edges *CD* and *DE* must be a part of every Hamilton circuit and that *CE* cannot be a part of any Hamilton circuit.

There are 2 possibilities at vertex *C*: edge *BC* or edge *CF*. Edge *BC* forces edge *FE* (otherwise there would be a circuit *E, B, C, D, E*) and edge *CF* forces edge *EB*.

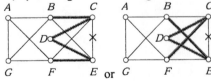

The first of these can be completed in 2 ways giving Hamilton circuits *A, B, C, D, E,*

F, G, A and *A, F, E, D, C, B, G, A* along with their mirror-image circuits *A, G, F, E, D, C, B, A* and *A, G, B, C, D, E, F, A*.

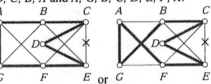

The second of these can also be completed in 2 ways giving Hamilton circuits *A, B, E, D, C, F, G, A* and *A, F, C, D, E, B, G, A* along with their mirror-image circuits *A, G, B, E, D, C, F, A* and *A, G, F, C, D, E, B, A*.

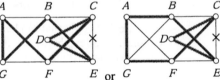

5. (a) *A, F, B, C, G, D, E*

(b) *A, F, B, C, G, D, E, A*

(c) *A, F, B, E, D, G, C*

(d) *F, A, B, E, D, C, G*

7. (a) 1. *A, B, C, D, E, F, A*
     2. *A, B, E, D, C, F, A*

Mirror-image circuits:
3. *A, F, E, D, C, B, A*
4. *A, F, C, D, E, B, A*

(b) 1. *D, E, F, A, B, C, D*
     2. *D, C, F, A, B, E, D*

Mirror-image circuits:
3. *D, C, B, A, F, E, D*
4. *D, E, B, A, F, C, D*

(c) The circuits in (b) are the same as the circuits in (a), just rewritten with a different starting vertex.

9. The degree of every vertex in a graph with a Hamilton circuit must be at least 2 since the circuit must "pass through" every vertex. This graph has 2 vertices of degree 1. Any Hamilton path passing through vertex *A* must contain edge *AB*. Any path passing through vertex *E* must contain edge *BE*. And, any path passing through vertex *C* must contain edge *BC*. Consequently, any path passing through vertices *A*, *C*, and *E* must contain at least three edges meeting at *B* and hence would pass through vertex *B* more than once. So, this graph does not have any Hamilton paths.

11. (a) *A, I, J, H, B, C, F, E, G, D*

   (b) *G, D, E, F, C, B, A, I, J, H*

   (c) If such a path were to start by heading left, it would not contain *C*, *D*, *E*, *F*, or *G* since it would need to pass through *B* again in order to do so. On the other hand, if a path were to start by heading right, it could not contain *A*, *I H*, or *J*.

   (d) Any circuit would need to cross the bridge *BC* twice (in order to return to where it started). But then *B* and *C* would be included twice.

13. (a) 6

   (b) *B, D, A, E, C, B*
       Weight = 6 + 1 + 9 + 4 + 7 = 27

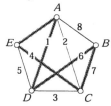

   (c) The mirror image *B, C, E, A, D, B*
       Weight = 7 + 4 + 9 + 1 + 6 = 27

15. (a) *A, D, F, E, B, C*
       Weight = 2 + 7 + 5 + 4 + 11 = 29

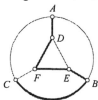

   (b) *A, B, E, D, F, C*
       Weight = 10 + 4 + 3 + 7 + 6 = 30

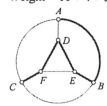

   (c) There are only two Hamilton paths that start at *A* and end at *C*. The path *A*, *D*, *F*, *E*, *B*, *C* found in part (a) is the optimal such path. It has weight 29.

B. Factorials and Complete Graphs

17. (a) $10! = 3,628,800$

   (b) $20! = 2,432,902,008,176,640,000$

   (c) $\dfrac{20!}{10!} = 670,442,572,800$

19. (a) $20! = 2,432,902,008,176,640,000$
          $\approx 2.43 \times 10^{18}$

   (b) $40! \approx 8.16 \times 10^{47}$

   (c) $(40-1)! = 39! \approx 2.04 \times 10^{46}$

21. (a) $10! = 10 \times 9 \times 8 \times \ldots \times 3 \times 2 \times 1$
          $= 10 \times 9!$
       So, $9! = \dfrac{10!}{10} = \dfrac{3,628,800}{10} = 362,880$.

(b) $11! = 11 \times 10 \times 9 \times \ldots \times 3 \times 2 \times 1 = 11 \times 10!$

So, $\dfrac{11!}{10!} = \dfrac{11 \times 10!}{10!} = 11$.

(c) $11! = 11 \times 10 \times (9 \times \ldots \times 3 \times 2 \times 1)$

$= 11 \times 10 \times 9!$

So, $\dfrac{11!}{9!} = \dfrac{11 \times 10 \times 9!}{9!} = 11 \times 10 = 110$.

(d) $\dfrac{101!}{99!} = \dfrac{101 \times 100 \times 99!}{99!}$

$= 101 \times 100$

$= 10,100$

23. (a) $\dfrac{9! + 11!}{10!} = \dfrac{9!}{10!} + \dfrac{11!}{10!}$

$= \dfrac{9!}{10 \times 9!} + \dfrac{11 \times 10!}{10!}$

$= \dfrac{1}{10} + 11$

$= 11.1$

(b) $\dfrac{101! + 99!}{100!} = \dfrac{101!}{100!} + \dfrac{99!}{100!}$

$= \dfrac{101 \times 100!}{100!} + \dfrac{99!}{100 \times 99!}$

$= 101 + \dfrac{1}{100}$

$= 101.01$

25. (a) $\dfrac{20 \times 19}{2} = 190$

(b) K_{21} has 20 more edges than K_{20}.

$\dfrac{21 \times 20}{2} = 210$

(c) If one vertex is added to K_{50} (making for 51 vertices), the new complete graph K_{51} has 50 additional edges (one to each old vertex). In short, K_{51} has 50

more edges than K_{50}. That is, $y - x = 50$.

27. (a) $120 = 5!$, so $N = 6$.

(b) $45 = \dfrac{9 \times 10}{2}$, so $N = 10$.

(c) $20,100 = \dfrac{200 \times 201}{2}$, so $N = 201$.

C. Brute Force and Nearest Neighbor Algorithms

29. (a)

Hamilton Circuit (Mirror Image)	Weight
A, B, C, D, A (A, D, C, B, A)	$38 + 22 + 8 + 12 = 80$
A, B, D, C, A (A, C, D, B, A)	$38 + 10 + 8 + 18 = 74$
A, C, B, D, A (A, D, B, C, A)	$18 + 22 + 10 + 12 = 62$

(b) A, D, C, B, A

Weight $= 12 + 8 + 22 + 38 = 80$

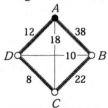

(c) B, D, C, A, B

Weight $= 10 + 8 + 18 + 38 = 74$

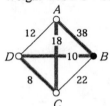

(d) C, D, B, A, C
Weight = $8 + 10 + 38 + 18 = 74$

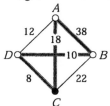

31. **(a)** B, C, A, E, D, B; Cost = $\$121 + \$119 + \$133 + \$199 + \$150 = \$722$

    (b) C, A, E, D, B, C; Cost = $\$119 + \$133 + \$199 + \$150 + \$121 = \$722$

    (c) D, B, C, A, E, D; Cost = $\$150 + \$121 + \$119 + \$133 + \$199 = \$722$

    (d) E, C, A, D, B, E; Cost = $\$120 + \$119 + \$152 + \$150 + \$200 = \$741$

33. **(a)** A, D, E, C, B, A
Cost of bus trip = $\$8 \times (185 + 302 + 165 + 305 + 500) = \$11,656$

    (b) A, D, B, C, E, A
Cost of bus trip = $\$8 \times (185 + 360 + 305 + 165 + 205) = \$9,760$

    (c) There are only 6 possible circuits that make B the first stop after A.

Hamilton Circuit	Cost
A, B, C, D, E, A	$\$8 \times (500 + 305 + 320 + 302 + 205) = \$13,056$
A, B, C, E, D, A	$\$8 \times (500 + 305 + 165 + 302 + 185) = \$11,656$
A, B, D, C, E, A	$\$8 \times (500 + 360 + 320 + 165 + 205) = \$12,400$
A, B, D, E, C, A	$\$8 \times (500 + 360 + 302 + 165 + 200) = \$12,216$
A, B, E, C, D, A	$\$8 \times (500 + 340 + 165 + 320 + 185) = \$12,080$
A, B, E, D, C, A	$\$8 \times (500 + 340 + 302 + 320 + 200) = \$13,296$

Optimal Hamilton circuit starting with A and then B: A, B, C, E, D, A
Cost of bus trip = $\$8 \times (500 + 305 + 165 + 302 + 185) = \$11,656$

35. **(a)** Atlanta, Columbus, Kansas City, Tulsa, Minneapolis, Pierre, Atlanta
Cost of this trip = $\$0.75 \times (533 + 656 + 248 + 695 + 394 + 1361) = \$2915.25$

    (b) The nearest-neighbor circuit with Kansas City as the starting vertex is Kansas City, Tulsa, Minneapolis, Pierre, Columbus, Atlanta, Kansas City
Written with starting city Atlanta, the circuit is Atlanta, Kansas City, Tulsa, Minneapolis, Pierre, Columbus, Atlanta
Cost of this trip = $\$0.75 \times (798 + 248 + 695 + 394 + 1071 + 533) = \$2804.25$

D.   Repetitive Nearest-Neighbor Algorithm

37.

Starting Vertex	Hamilton Circuit	Weight of Circuit
A	A, D, E, B, C, A	$2.1 + 1.2 + 2.8 + 2.6 + 2.3 = 11$
B	B, A, D, E, C, B	$2.2 + 2.1 + 1.2 + 3.1 + 2.6 = 11.2$
C	C, D, E, A, B, C	$1.4 + 1.2 + 2.4 + 2.2 + 2.6 = 9.8$
D	D, E, A, B, C, D	$1.2 + 2.4 + 2.2 + 2.6 + 1.4 = 9.8$
E	E, D, C, A, B, E	$1.2 + 1.4 + 2.3 + 2.2 + 2.8 = 9.9$

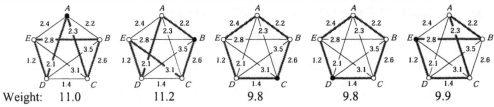

Weight: 11.0   11.2   9.8   9.8   9.9

Starting with vertex B, the shortest circuit is B, C, D, E, A, B or B, C, D, E, A, B with weight 9.8.

39.

Starting Vertex	Hamilton Circuit	Length of Circuit (Miles)
A	A, D, B, C, E, A	$185 + 360 + 305 + 165 + 205 = 1220$
B	B, C, E, A, D, B	$305 + 165 + 205 + 185 + 360 = 1220$
C	C, E, A, D, B, C	$165 + 205 + 185 + 360 + 305 = 1220$
D	D, A, C, E, B, D	$185 + 200 + 165 + 340 + 360 = 1250$
E	E, C, A, D, B, E	$165 + 200 + 185 + 360 + 340 = 1250$

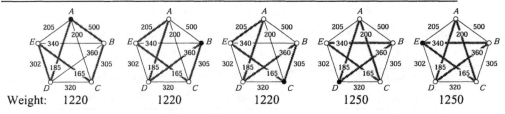

Weight: 1220   1220   1220   1250   1250

The shortest circuit (written as starting and ending at A) is A, D, B, C, E, A with length 1220. The cost of this bus trip is $\$8 \times 1220 = \$9,760$.

41.	Starting Vertex	Hamilton Circuit	Length of Circuit (Miles)
	A	A, C, K, T, M, P, A	$533 + 656 + 248 + 695 + 394 + 1361 = 3887$
	C	C, A, T, K, M, P, C	$533 + 772 + 248 + 447 + 394 + 1071 = 3465$
	K	K, T, M, P, C, A, K	$248 + 695 + 394 + 1071 + 533 + 798 = 3739$
	M	M, P, K, T, A, C, M	$394 + 592 + 248 + 772 + 533 + 713 = 3252$
	P	P, M, K, T, A, C, P	$394 + 447 + 248 + 772 + 533 + 1071 = 3465$
	T	T, K, M, P, C, A, T	$248 + 447 + 394 + 1071 + 533 + 772 = 3465$

The shortest circuit is M, P, K, T, A, C, M. Written starting from Atlanta, this is Atlanta, Columbus, Minneapolis, Pierre, Kansas City, Tulsa, Atlanta and has a length of 3252 miles.

E.   Cheapest-Link Algorithm

43. The successive steps are shown in the following figures.

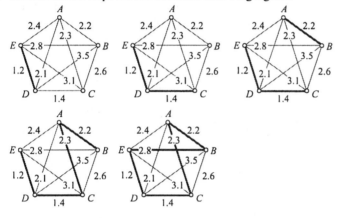

After the fourth step, the next cheapest edge is 2.4. Since this edge makes a circuit, we skip this edge and try the next cheapest edge. The next cheapest edge is 2.6 but that makes three edges come together at vertex C so we skip this edge and choose the next cheapest edge which is 2.8. This edge completes the Hamilton circuit. The shortest Hamilton circuit found using this algorithm is B, E, D, C, A, B. The weight of this circuit is $1.2 + 1.4 + 2.2 + 2.3 + 2.8 = 9.9$.

45. The successive steps are shown in the following figures.

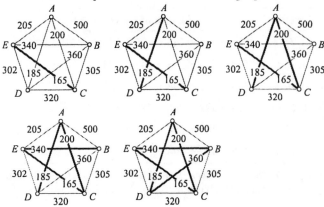

After the third step, the next cheapest edge is 205. But that edge would make three edges come together at vertex *A* so we skip it and move to the next cheapest edge which is 302. Since this edge makes a circuit, we skip this edge and try the next cheapest edge. The next cheapest edge is 305 but that makes three edges come together at vertex *C* so we skip this edge and choose the next cheapest edge which is 320. That edge makes a circuit and also makes three edges come together at vertex *C*. So, we skip it and choose the next cheapest edge which is 340.

The shortest Hamilton circuit found using this algorithm is *B, E, C, A, D, B*. The cost of this bus trip is $8 \times (165 + 185 + 200 + 340 + 360) = \$10,000$.

47.

Link Added to Circuit	Cost of Link (Miles)
Kansas City - Tulsa	248
Pierre - Minneapolis	394
Minneapolis - Kansas City	447
Atlanta - Columbus	533
Atlanta - Tulsa	772
Columbus - Pierre	1071

The circuit is Atlanta, Columbus, Pierre, Minneapolis, Kansas City, Tulsa, Atlanta.
Total mileage = 248 + 394 + 447 + 533 + 772 + 1071 = 3465 miles.

F.   Miscellaneous

49. (a) *A, B, C, D, E, A*; Weight = 1 + 1 + 1 + 1 + 100 = 104

   (b) *A, B, C, D, E, A*; Weight = 1 + 1 + 1 + 1 + 100 = 104

   (c) *A, C, D, E, B, A*; Weight = 2 + 1 + 1 + 2 + 1 = 7

51. (a)

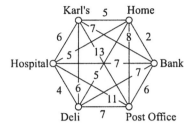

(b) Just "eyeballing it" will give the optimal circuit in this case. It is: Home, Bank, Post Office, Deli, Hospital, Karl's , Home. The total length of the trip is $2 + 6 + 7 + 4 + 6 + 5 = 30$ miles.

53. (a)

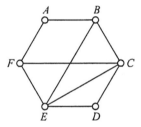

(b) Any Hamilton circuit gives a possible seating arrangement. One possibility is A, B, C, D, E, F.

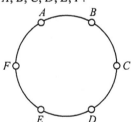

Seating Arrangement

(c) Yes, there is. It corresponds to the Hamilton circuit A, B, E, D, C, F, A.

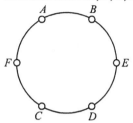

Seating Arrangement

55. If we draw the graph describing the friendships among the guests (see figure) we can see that the graph does not have a Hamilton circuit, which means it is impossible to seat everyone around the table with friends on both sides.

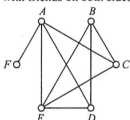

JOGGING

57.

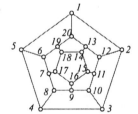

59. $A, B, C, D, J, I, F, G, E, H$

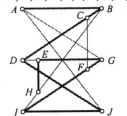

61. The 2 by 2 grid graph cannot have a Hamilton circuit because each of the 4 corner vertices as well as the interior vertex I must be preceded and followed by a boundary vertex. But there are only 4 boundary vertices–not enough to go around.

63. **(a)**

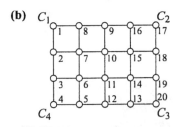

(b)

(c) Think of the vertices of the graph as being colored like a checker board with C_1 being a red vertex. Then each time we move from one vertex to the next we must move from a red vertex to a black vertex or from a black vertex to a red vertex. Since there are 10 red vertices and 10 black vertices and we are starting with a red vertex, we must end at a black vertex. But C_2 is a red vertex. Therefore, no such Hamilton path is possible.

65.

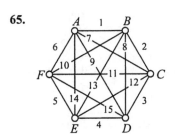

The optimal Hamilton circuit is A, B, C, D, E, F, A. The edges in this circuit are the edges with the 6 lowest weights.

67. **(a)** Any circuit would need to cross the bridge twice (in order to return to where it started). But then each vertex at the end of the bridge would be included twice.

(b) The following graph has a Hamilton path from B to C.

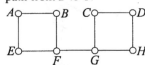

69. Dallas, Houston, Memphis, Louisville, Columbus, Chicago, Kansas City, Denver, Atlanta, Buffalo, Boston, Dallas.

The process starts by finding the smallest number in the Dallas row. This is 243 miles to Houston. We cross out the Dallas column and proceed by finding the smallest number in the Houston row (other than that representing Dallas which has been crossed out). This is 561 miles to Memphis. We now cross out the Houston column and continue by finding the smallest number in the Memphis row (other than those representing Dallas and Houston which have been crossed out). The process continues in this fashion until all cities have been reached.

71. **(a)** Julie should fly to Detroit. The optimal route will then be to drive $68 + 56 + 68$

+ 233 + 78 + 164 + 67 + 55 = 789 miles: Detroit-Flint-Lansing-Grand Rapids-Cheboygan-Sault Ste. Marie-Marquette-Escanaba-Menominee for a total cost of 789 × ($0.39) + $2.50 = $310.21. Since Julie can drive from Menominee back to Detroit (via Sault Ste. Marie) in a matter of 227 + 78 + 280 = 585 miles at a cost of 585 × ($0.39) + $2.50 = $230.65, she should do so and drop the rental car back in Detroit (assuming she need not pay extra for gas!). The total cost of her trip would then be $310.21 + $230.65 = $540.86.

(b) The optimal route would be for Julie to fly to Detroit and drive 68 + 56 + 68 + 233 + 78 + 164 + 67 + 55 = 789 miles along the Hamilton path Detroit-Flint-Lansing-Grand Rapids-Cheboygan-Sault Ste. Marie-Marquette-Escanaba-Menominee for a total cost of 789 × ($0.49) + $2.50 = $389.11.

Mini-Excursion 2

WALKING

A. Graph Colorings and Chromatic Numbers

1. (a) Many answers are possible.

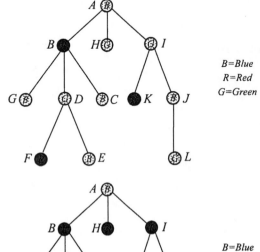

B=Blue
R=Red
G=Green

(b)

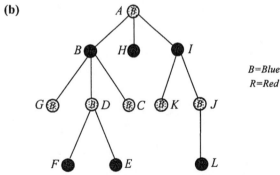

B=Blue
R=Red

(c) $\chi(G) = 2$. At least 2 colors are needed and (b) shows that G can be colored with 2 colors.

3. (a) Many answers are possible.

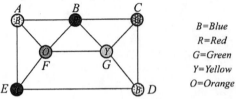

B=Blue
R=Red
G=Green
Y=Yellow
O=Orange

130

(b)

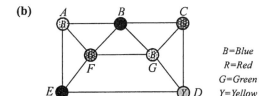

B=Blue
R=Red
G=Green
Y=Yellow

(c) $\chi(G) = 4$. The graph cannot be colored with 3 colors. If we try to color G with 3 colors and start with a triangle, say *AEF*, and color it blue, red, green, then B is forced to be red, G is forced to be blue, C is forced to be green, and then D will require a fourth color.

5. (a)

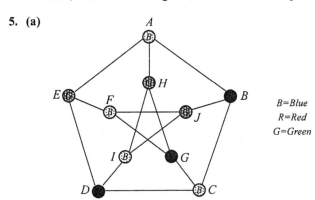

B=Blue
R=Red
G=Green

(b) $\chi(G) = 3$. Three colors are needed because A, B, C, D, and E form a circuit of length 5.

7. (a) Every vertex of K_n is adjacent to every other vertex, so every vertex has to be colored with a different color.

(b) $\chi(G) = n - 1$. If we remove one edge from K_n, then there are two vertices that are not adjacent. They can be colored with the same color (say blue). The remaining vertices have to be colored with different colors other than blue.

9. (a) Adjacent vertices around the circuit can alternate colors (blue, red, blue, red, ...).

(b) To color an odd circuit we start by alternating two colors (blue, red, blue, red,...), but when we get to the last vertex, it is adjacent to both a blue and a red vertex, so a third color is needed.

11. $\chi(G) = 2$.

Since a tree has no circuits, we can start with any vertex v, color it blue and alternate blue, red, blue, red, blue, ... along any path of the tree. Every vertex of the tree is in a unique path joining it to v and can be colored either red or blue.

B.   Map Coloring

13.   Answers may vary. One possible answer is shown below.

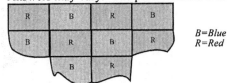

B=Blue
R=Red

15.   (a) & (b)

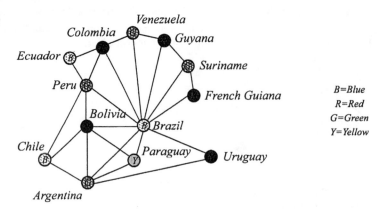

B=Blue
R=Red
G=Green
Y=Yellow

List of vertices (by decreasing order of degrees): Brazil (10), Bolivia (5), Argentina (5), Peru (5), Columbia (4), Chile (3), Paraguay (3), Venezuela (3), Guyana (3), Suriname (3), Ecuador (2), French Guiana (2), Uruguay (2).

Priority list of colors: Blue, Red, Green, Yellow.

(c)   The chromatic number is 4 since Brazil, Bolivia, Argentina and Paraguay are all adjacent to each other.

C.   Miscellaneous

17.   (a)   All the vertices in A can be colored with the same color, and all of the vertices in B can be colored with a second color.

(b)   Suppose G has $\chi(G) = 2$. Say the vertices are colored blue and red. Then let A denote the set of blue vertices and B the set of red vertices.

(c)   If G had a circuit with an odd number of vertices then that circuit alone would require 3 color (see Exercise 9). But from (a) we know that $\chi(G) = 2$.

19.   (a)   Since the sum of the degrees of all the vertices must be even (see Euler's Sum of Degrees theorem in Chapter 5), it follows that the number of vertices of degree 3 must be even, and thus, n must be odd. If $n = 3$, there can be no vertices of degree 3 unless there are multiple edges.

(b) From the strong version of Brook's theorem we have $\chi(G) \leq 3$. We know $\chi(G)$ cannot be 1 (see Exercise 8). Moreover, G cannot be a bipartite graph (see next paragraph), so $\chi(G)$ cannot be 2 (see Exercise 17). It follows that $\chi(G) = 3$.

A graph with *n-1* vertices of degree 3 and one vertex of degree 2 cannot be bipartite because the vertex of degree 2 must be in one of the two parts, say *A*. Then the number of edges coming out of *A* is 2 plus a multiple of 3. On the other hand, the number of edges coming out of *B* is a multiple of 3. But in a bipartite graph the number of edges coming out of part *A* must equal the number of edges coming out of part *B* (they are both equal to the total number of edges in the graph).

21. Suppose that the graph *G* is colored with $\chi(G)$ colors: Color 1, Color 2, ..., Color *K* (for simplicity we will use *K* for $\chi(G)$). Now make a list $v_1, v_2, \ldots, v_n$ of the vertices of the graph as follows: All the vertices of Color 1 (in any order) are listed first (call these vertices Group 1), the vertices of Color 2 are listed next (call these vertices Group 2), and so on, with the vertices of Color *K* listed last (Group *K*). Now when we apply the greedy algorithm to this particular list, the vertices in Group 1 get Color 1, the vertices in Group 2 get either Color 1 or Color 2, the vertices in Group 3 get either Color 1, or Color 2, or at worst, Color 3, and so on. The vertices in Group *K* get Color 1, or 2, ..., or at worst, Color *K*. It follows that the greedy algorithm gives us an optimal coloring of the graph.

23. The solution appears in Mini-Excursion 2 after the References and Further Readings.

Chapter 7

A.   Trees

1. **(a)** Is a tree.

   (b) Is not a tree (is not connected).

   (c) Is not a tree (has a circuit).

   (d) Is a tree.

3. **(a)** (II) A tree with 8 vertices must have 7 edges.

   (b) (III)

   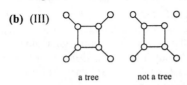

   a tree   not a tree

   (c) (I) If every edge of a graph is a bridge, then the graph must be a tree (Property 2).

5. **(a)** (I) If there is exactly one path joining any two vertices of a graph, the graph must be a tree.

   (b) (II) A tree with 8 vertices must have 7 edges and every edge must be a bridge.

   (c) (I) If every edge is a bridge, then the graph has no circuits. Since the graph is also connected, it must be a tree.

7. **(a)** (III)

   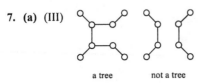

   a tree   not a tree

   (b) (II) A tree has no circuits.

   (c) (I) A graph with 8 vertices, 7 edges, and no circuits, must also be connected and hence must be a tree.

9. **(a)** (II) Since the degree of each vertex is even, it must be at least 2. Thus, the sum of the degrees of all 8 vertices must be at least 16. But, a tree with 8 vertices must have 7 edges, and the sum of the degrees of all the vertices would have to be 14.

   (b) (III)

   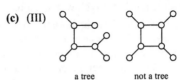

   a tree   not a tree

   (c) (III)

   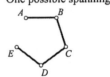

   a tree   not a tree

B.   Spanning Trees

11. **(a)** One possible spanning tree is

   (b) The only spanning tree is

   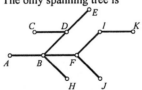

   (c) One possible spanning tree is

134

13. (a) There are three ways to eliminate the only circuit *A*, *B*, *C*, *A*.

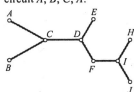

(b) There are five ways to eliminate the only circuit *B*, *C*, *D*, *E*, *F*, *B*.

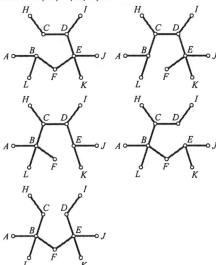

15. (a) Each spanning tree excludes one of the edges *AB*, *BC*, *CA* and one of the edges *DE*, *EI*, *IF*, *FD* so there are $3 \times 4 = 12$ different spanning trees.

(b) Each spanning tree excludes one of the edges *EJ*, *JK*, *KE* and one of the edges *BC*, *CD*, *DE*, *EF*, *FB* so there are $3 \times 5 = 15$ different spanning trees.

17. (a) Each spanning tree excludes one of the edges *AB*, *BC*, *CA*, one of the edges *DE*, *EI*, *IF*, *FD*, and one of the edges *HI*, *IJ*, *JH*, so there are $3 \times 4 \times 3 = 36$ different spanning trees.

(b) Each spanning tree excludes one of the edges *AB*, *BL*, *LA*, one of the edges *BC*, *CD*, *DE*, *EF*, *FB*, and one of the edges *EJ*, *JK*, *KE*, so there are $3 \times 5 \times 3 = 45$ different spanning trees.

C. Minimum Spanning Trees and Kruskal's Algorithm

19. (a) Add edges to the tree in the following order: *EC*, *AD*, *AC*, *BC*.

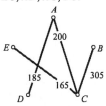

(b) The total weight is $165 + 185 + 200 + 305 = 855$.

21. (a) Add edges to the tree in the following order: *DC*, *EF*, *EC*, *AB*, *AC*.

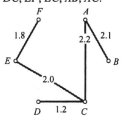

(b) The total weight is $1.2 + 1.8 + 2.0 + 2.1 + 2.2 = 9.3$.

23. Add edges to the tree in the following order: Kansas City – Tulsa (248), Pierre – Minneapolis (394), Minneapolis – Kansas City (447), Atlanta – Columbus (533), Columbus – Kansas City (656).

25. A spanning tree for the 20 vertices will have exactly 19 edges and so the cost will be

$$19 \text{ edges} \times \frac{1 \text{ mile}}{2 \text{ edges}} \times \frac{\$40,000}{1 \text{ mile}} = \$380,000.$$

D.   Steiner Points and Shortest Networks

27. (a) The measure of angle A is $180° - 24° - 27° = 129°$. Since this is at least 120°, there is no Steiner point. Therefore, the shortest network is the minimum spanning tree, which consists of the edges AB and AC. The length is 270 miles + 310 miles = 580 miles.

(b) The measure of angle A is $180° - 33° - 27° = 120°$. Since this is at least 120° (in fact, exactly that), there is no Steiner point. Therefore, the shortest network is the minimum spanning tree, which consists of the edges AB and AC. The length is 212 miles + 173 miles = 385 miles.

29. (a) The measure of angle A is $180° - 20° - 20° = 140°$. Since this is at least 120°, there is no Steiner point. Therefore, the shortest network is the minimum spanning tree. We can deduce that the length of side AC is 183 km because the measures of angles B and C are equal and so the lengths of the sides opposite those angles must be equal. Thus, the minimum spanning tree consists of the edges AB and AC, and the length is 183 km + 183 km = 366 km.

(b) Since side AC is shorter than side AB, we can deduce that the measure of angle B is less than the measure of angle C, or less than 28°. So the measure of angle A is at least $180° - 28° - 28° = 124°$. Since this is at least 120°, there is no Steiner point. Therefore, the shortest network is the minimum spanning tree, which consists of the edges AB and AC. (Edge BC must be longer than either AB or AC since it is opposite the largest angle.) The length is 181 km + 153 km = 334 km.

31. Since the measure of angle CAB is 120°, there is no Steiner point. Therefore, the shortest network is the minimum spanning tree, which consists of the edges AB and AC. In a 30-60-90 triangle, the hypotenuse is twice as long as the shortest leg (see Exercise 45(a)). Since the shortest leg of

both of the 30-60-90 triangles in the figure is 85 miles, the hypotenuse for each 30-60-90 triangle is 170 miles. That is, $AB = 170$ miles and $AC = 170$ miles. The length of the shortest network is 170 miles + 170 miles = 340 miles.

33. The sum of the distances from Z to A, B, and C is 232 miles, the sum of the distances from X to A, B, and C is 240 miles, and the sum of the distances from Y to A, B, and C is 243 miles. Since one of the points is the Steiner point and the Steiner point is the point that makes the shortest network, Z is the Steiner point.

35. (a) $CE + ED + EB$ is larger since $CD + DB$ is the shortest network connecting the cities C, D, and B.

(b) $CD + DB$ is the shortest network connecting the cities C, D, and B since angle CDB is 120° and so the shortest network is the same as the minimum spanning tree.

(c) $CE + EB$ is the shortest network connecting the cities C, E, and B since angle CEB is more than 120° and so the shortest network is the same as the minimum spanning tree.

E.   Miscellaneous

37. (a) $k = 5, 2, 1, 0$ are all possible. The network could have no circuits (k=5),

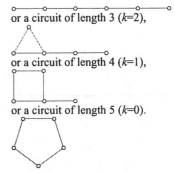

or a circuit of length 3 (k=2),

or a circuit of length 4 (k=1),

or a circuit of length 5 (k=0).

(b) Using the same pattern as in (a), values of $k = 123$ and $0 \le k \le 120$ are all possible.

39. (a) Since $M = N - 1$, the network is a tree by Property 4. So, there are no circuits in the network.

(b) The redundancy of this network is $R = 1$ so one edge will need to be "discarded" in order to form a spanning tree. This means that there is one circuit in the network.

(c) 118. Each edge other than those in the circuit of length 5 is a bridge.

41. (a) Note that the graph is not connected.

(b) The graph is, again, not connected.

(c) Other examples are possible.

43. Since S is a Steiner point, the measure of angle ASB is 120°. Since the triangle is isosceles, the measure of angle ASB is the same as the measure of angle BSC (half of 90°). That is, the measure of angle ASB is 45°. So, the measure of angle BAS is $180° - 120° - 45° = 15°$.

45. (a) Imagine that the figure is the top half of an equilateral triangle. Then, it becomes clear that h is twice the length of s. That is, $h = 2s$.

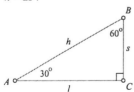

(b) By the Pythagorean theorem, $h^2 = s^2 + l^2$. Then, using the result of (a), we have $(2s)^2 = s^2 + l^2$ so that $4s^2 = s^2 + l^2$ and $3s^2 = l^2$. Since $l > 0$, it follows that $l = \sqrt{3}s$.

(c) $h = 2s = 2(21.0 \text{ cm}) = 42 \text{ cm}$
$l = \sqrt{3}s = \sqrt{3}(21.0 \text{ cm}) \approx 36.3 \text{ cm}$

(d) $s = \dfrac{1}{2}h = \dfrac{1}{2}(30.4 \text{ cm}) = 15.2 \text{ cm}$
$l = \sqrt{3}s = \sqrt{3}(15.2 \text{ cm}) \approx 26.3 \text{ cm}$

47. In a 30°-60°-90° triangle, the side opposite the 30° angle is one half the length of the hypotenuse and the side opposite the 60° angle is $\dfrac{\sqrt{3}}{2}$ times the length of the hypotenuse (See Exercise 45). Therefore, the distance from C to J is $\dfrac{\sqrt{3}}{2} \times 500$ miles ≈ 433 miles and so the total length of the T-network (rounded to the nearest mile) is
433 miles + 500 miles = 933 miles.

JOGGING

49. (a) Each spanning tree excludes one of the edges AB, BC, CA, and one of the edges HI, IJ, JH. Since the circuits C, D, E, C and D, E, I, F, D share a common edge there are two ways to exclude edges to form a spanning tree. If one of the excluded edges is the common edge DE, then any one of the other 5 edges $CD, CE, EI, DF,$ or FI could be excluded to form a spanning tree. If, on the other hand, one of the excluded edges is not the common edge DE, then one excluded edge has to be either CE or CD and the other excluded edge must be either $EI, DF,$ or FI. Thus, there are $3 \times (5 + 2 \times 3) \times 3 = 99$ different spanning trees.

(b) Each spanning tree excludes one of the edges AB, BL, LA, and one of the edges EJ, JK, KE. Since the circuits C, D, I, H, C and B, C, D, E, F, B share a common edge there are two ways to exclude edges to form a spanning tree. If one of the excluded edges is the common edge CD, then any one of the other 7 edges $CH, HI, ID, BC, BF, EF,$ or DE could be excluded to form a spanning tree. Or, if one of the excluded edges is not the common edge CD, then one excluded edge has to be either CH, HI or ID and the other excluded edge must be either $BC, BF, EF,$ or DE. Thus, this network has

$3 \times (7 + 3 \times 4) \times 3 = 171$ different spanning trees.

51. (a) A regular tree with $N = 2$ vertices:

(b) Suppose that a tree is regular. Since two vertices of a tree must have degree 1 (see Exercise 50(b)), every vertex must have degree 1. So the sum of the degrees is N. However, the sum of the degrees is also twice the number of edges. So, there are $N/2$ edges. In order to be a tree, it must be that $N-1 = N/2$ (the number of edges must be one fewer than the number of vertices). Solving this equation gives $N = 2$ as the only solution. So, if $N \geq 3$, the graph cannot be regular.

53. (a) If $R = 0$, then $M = N-1$ (the number of edges is one less than the number of vertices). So, the network is a tree.

(b) If $R = 1$, then $M = N$ (the number of edges is the same as the number of vertices). So, the network is not a tree. From the definition of a tree, there must be at least one circuit. Suppose that the graph had 2 (or more) circuits. Then, there would be an edge of one of the circuits that is not an edge of another circuit. Removing such an edge would leave us with a connected graph with the number of vertices being one more than the number of edges (i.e. a tree). But this tree would have a circuit. This is impossible, so there cannot be more than 1 circuit.

(c) The maximum redundancy occurs when the degree of each vertex is $N-1$. In that case (a complete graph), the number of edges is $(N-1) + (N-2) + \ldots + 3 + 2 + 1$. So, $R = (N-1) + (N-2) + \ldots + 3 + 2 + 1 - (N-1) = (N-2) + \ldots + 3 + 2 + 1 = \left(N^2 - 3N + 2\right)/2$. [See chapter 1 or chapter 10 on evaluating this sum.]

55. (a) According to Cayley's theorem, there are $3^{3-2} = 3$ spanning trees in a complete graph with 3 vertices, which is confirmed by the following figures.

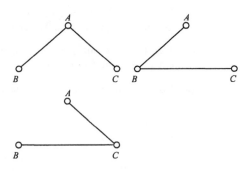

Likewise, for the completed graph with 4 vertices, Cayley's theorem predicts $4^{4-2} = 16$ spanning trees, which is confirmed by the following figures.

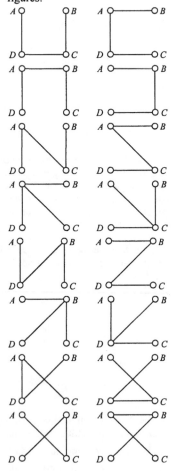

(b) The complete graph with N vertices has $(N-1)!$ Hamilton circuits and N^{N-2} spanning trees. Since $(N-1)! = 2 \times 3 \times 4 \times ... \times (N-1)$ and $N^{N-2} = N \times N \times N \times ... \times N$, and both expressions have the same number of factors, with each factor in $(N-1)!$ smaller than the corresponding factor in N^{N-2}, we know that for $N = 3$, $(N-1)! < N^{N-2}$. Thus, the number of spanning trees is larger than the number of Hamilton circuits.

57.

Edges in Spanning Tree	Junction Points	Total Cost
AB, AC, AD	*A*	$85 + 50 + 45 + 25 = 205$
AB, BC, BD	*B*	$85 + 90 + 75 + 5 = 255$
AC, BC, CD	*C*	$50 + 90 + 70 + 15 = 225$
AD, BD, CD	*D*	$45 + 75 + 70 + 20 = 210$
AD, AB, BC	*A, B*	$45 + 85 + 90 + 25 + 5 = 250$
AB, AC, BD	*A, B*	$85 + 50 + 75 + 25 + 5 = 240$
AB, BC, CD	*B, C*	$85 + 90 + 70 + 5 + 15 = 265$
BD, BC, AC	*B, C*	$75 + 90 + 50 + 5 + 15 = 235$
BC, CD, AD	*C, D*	$90 + 70 + 45 + 15 + 20 = 240$
AC, CD, BD	*C, D*	$50 + 70 + 75 + 15 + 20 = 230$
AB, AD, CD	*A, D*	$85 + 45 + 70 + 25 + 20 = 245$
AC, AD, BD	*A, D*	$50 + 45 + 75 + 25 + 20 = 215$
AD, AC, BC	*A, C*	$45 + 50 + 90 + 25 + 15 = 225$
AB, BD, CD	*B, D*	$85 + 75 + 70 + 5 + 20 = 255$
AD, BD, BC	*B, D*	$45 + 75 + 90 + 20 + 5 = 235$
AB, AC, CD	*A, C*	$85 + 50 + 70 + 25 + 15 = 245$

The minimum cost network connecting the 4 cities has a 3-way junction point at *A* and has a total cost of 205 million dollars.

59. (a) The switching station should be located 1 mile north of the airport. [Most of the options can be eliminated by common sense. Since there are only a handful of viable options, trial and error will discover this solution.]

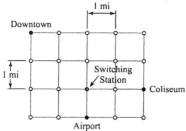

(b) Since the optimal network requires 7 miles of track and 1 switching station, the cost of the network is $7.5 million.

61. (a) $m(\angle BFA) = 120°$ since $m(\angle EFG) = 60°$ (supplementary angles). Similarly, $m(\angle AEC) = 120°$ and $m(\angle CGB) = 120°$.

(b) In $\triangle ABF$, $m(\angle BFA) = 120°$ [from part (a)] and so $m(\angle BAF) + m(\angle ABF) = 60°$. It follows therefore that $m(\angle BAF) < 60°$ and $m(\angle ABF) < 60°$. Similarly, $m(\angle ACE) < 60°$, $m(\angle CAE) < 60°$, $m(\angle BCG) < 60°$, and $m(\angle CBG) < 60°$. Consequently, $m(\angle A) = m(BAF) + m(CAE) < 60° + 60° = 120°$. Likewise, $m(\angle B) < 120°$ and $m(\angle C) < 120°$. So, triangle ABC has a Steiner point S.

(c) Any point X inside or on $\triangle ABF$ (except vertex F) will have $m(\angle AXB) > 120°$. Any point X inside or on $\triangle ACE$ (except vertex E) will have $m(\angle AXC) > 120°$. Any point X inside or on $\triangle BCG$ (except vertex G) will have $m(\angle BXC) > 120°$. If S is the Steiner point, $m(\angle ASB) = m(\angle ASC) = m(BSC) = 120°$, and so S cannot be inside or on $\triangle ABF$ or $\triangle ACE$ or $\triangle BCG$. It follows that the Steiner point S must lie inside $\triangle EFG$.

63. The length of the network is $4x + (500 - x) = 3x + 500$, where $250^2 + \left(\dfrac{x}{2}\right)^2 = x^2$ (see figure). Rewriting the equation gives $250{,}000 + x^2 = 4x^2$ or $3x^2 = 250{,}000$. Solving this gives $x = \sqrt{\dfrac{250{,}000}{3}} = \dfrac{500}{\sqrt{3}} = \dfrac{500\sqrt{3}}{3}$, and so the length of the network is $3\dfrac{500\sqrt{3}}{3} + 500 = 500\sqrt{3} + 500 \approx 1366$ miles.

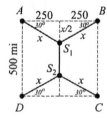

65. (a) There will be two subtrees formed as shown below.

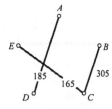

(b) The tree formed by Boruvka's algorithm is shown below.

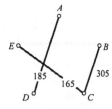

Chapter 8

WALKING

A.   Directed Graphs

1. (a) Vertex set: V = {A, B, C, D, E}; Arc set: A = {AB, AC, BD, CA, CD, CE, EA, ED}

   (b) indeg(A) = 2, indeg(B) = 1, indeg(C) = 1, indeg(D) = 3, indeg(E) = 1.

   (c) outdeg(A) = 2, outdeg(B) = 1, outdeg(C) = 3, outdeg(D) = 0, outdeg(E) = 2.

3. (a) C and E

   (b) B and C

   (c) B, C, and E

   (d) No vertices are incident from D.

   (e) CD, CE and CA

   (f) No arcs are adjacent to CD.

5. (a) 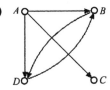   **(b)**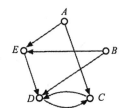

7. (a) 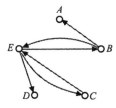   **(b)**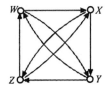

9. (a) 2 (these correspond to arcs AB and AE)

   (b) 1 (this corresponds to arc EA)

   (c) 1 (this corresponds to arc DB)

   (d) 0 (there are no arcs incident to D)

11. **(a)** A, B, D, E, F is one possible path.

    (b) A, B, D, E, C, F

    (c) B, D, E, B

    (d) The outdegree of vertex F is 0, so it cannot be part of a cycle.

    (e) The indegree of vertex A is 0, so it cannot be part of a cycle.

    (f) B, D, E, B is the only cycle

13.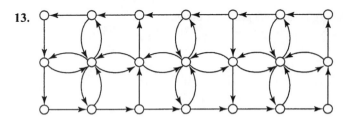

15. **(a)** B, since B is the only person that everyone respects.

    (b) A, since A is the only person that no one respects.

B.   Project Digraphs

17.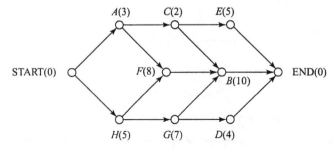

19.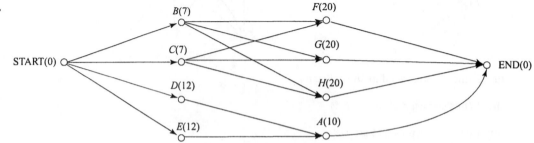

21.

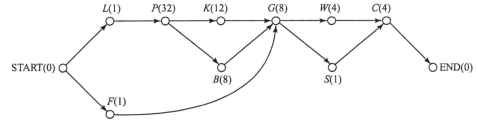

C.   Schedules and Priority Lists

23. **(a)** There are $31 \times 3 = 93$ processor hours available. Of these,
$10 + 7 + 11 + 8 + 9 + 5 + 3 + 6 + 4 + 7 + 5 = 75$ hours are used. So there must be
$93 - 75 = 18$ hours idle time.

    (b) There is a total of 75 hours of work to be done. Three processors working without any idle time
would take $\dfrac{75}{3} = 25$ hours to complete the project.

25. There is a total of 75 hours of work to be done. Dividing the work equally between the six processors
would require each processor to do $\dfrac{75}{6} = 12.5$ hours of work. Since there are no $\dfrac{1}{2}$-hour jobs, the
completion time could not be less than 13 hours.

27.

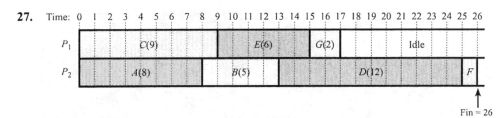

29.

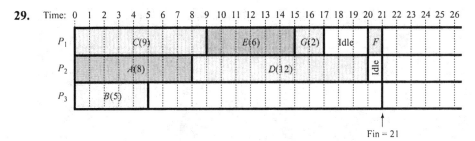

31. There are many different priority lists that will produce the given schedule. Here are five:
B,C,A,E,D,G,F; B,C,A,E,D,F,G; B,C,A,E,G,F,D; B,C,A,E,G,D,F; B,C,A,D,E,G,F.

33. According to the precedence relations, *G* cannot be started until *K* is completed.

35. Project Digraph:

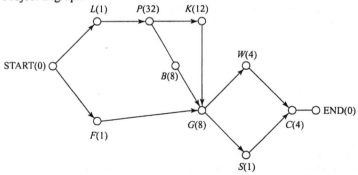

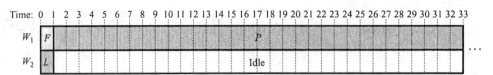

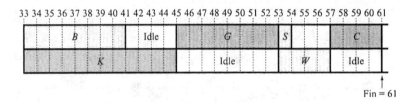

D. The Decreasing-Time Algorithm

37. Decreasing-Time List: *D, C, A, E, B, G, F*; so the schedule is:

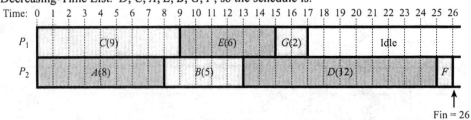

39. Refer to the project digraph from problem 35 to create the schedule. The Decreasing-Time List is *P, K, B, G, C, W, F, L, S.* The finishing time is *T* = 61.

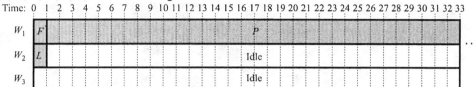

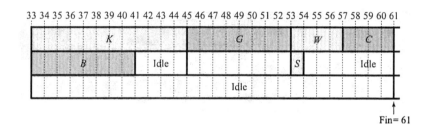

Fin = 61

41. (a) The 13 jobs *A–M* are already listed in decreasing order. So, the decreasing-time list is: *A, B, C, D, E, F, G, H, I, J, K, L, M* and (since the precedence relations play no role in this problem) the schedule is:

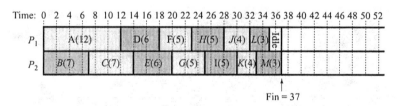

Fin = 37

(b)

Time: 0 2 4 6 8 10 12 14 16 18 20 22 24 26 28 30 32 34 36 38 40 42 44 46 48 50 52

| P_1 | A(12) | D(6) | F(5) | H(5) | J(4) | K(4) |
| P_2 | B(7) | C(7) | E(6) | G(5) | I(5) | L(3) | M(3) |

Opt = 36

43. (a) The 13 jobs *A–M* are already listed in decreasing order. So, the decreasing-time list is: *A, B, C, D, E, F, G, H, I, J, K, L, M* and since each task of length 5 cannot start until tasks *B* and *C* are completed the schedule is:

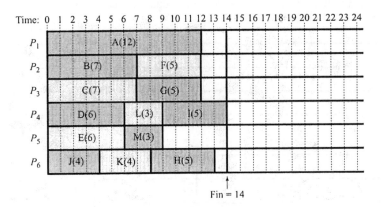

Fin = 14

(b)

Opt = 12

(c) For six copiers, the optimal completion time is 12 hours and one of the tasks takes 12 hours, thus the job cannot be completed in less than 12 hours.

E.   Critical Paths and Critical-Path Algorithm

45. (a)

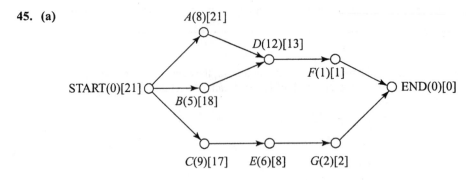

(b) Start, *A*, *D*, *F*, END is the critical path, with a critical time of 21.

(c) The critical path list is: $A[21]$, $B[18]$, $C[17]$, $D[13]$, $E[8]$, $G[2]$, $F[1]$.
The schedule is:

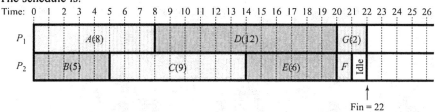

(d) There are a total of 43 work units, so the shortest time the project can be completed by 2 workers is $\frac{43}{2} = 21.5$ time units. Since there are no tasks with less than 1 time unit, the shortest time the project can actually be completed is 22 hours.

47. The critical path list is: $B[46]$, $A[44]$, $E[42]$, $D[39]$, $F[38]$, $I[35]$, $C[34]$, $G[24]$, $K[20]$, $H[10]$, $J[5]$.
The schedule is:

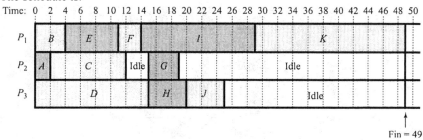

49. The project digraph, with critical times is:

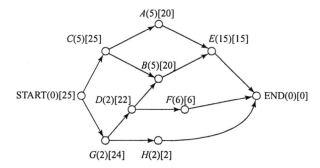

From this, we see the critical path list is: $C[25]$, $G[24]$, $D[22]$, $A[20]$, $B[20]$, $E[15]$, $F[6]$, $H[2]$.
The schedule is:

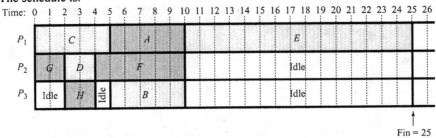

Fin = 25

F.　Scheduling with Independent Tasks

51. (a)

Fin = 18

(b) The optimal finishing time for $N = 2$ processors is $Opt = 18$ as shown in (a).

(c) The relative error expressed as a percent is $\varepsilon = \dfrac{Fin - Opt}{Opt} = \dfrac{18 - 18}{18} = 0\%$.

53. (a)

Fin = 7

(b) The optimal finishing time for $N = 3$ processors is $Opt = 7$ as shown in (a).

55. (a) Since all tasks are independent, the critical path list is identical to a decreasing-time list. The critical path list is: $E, I, D, H, C, G, A, B, F$.

The schedule using $N = 4$ processors is:

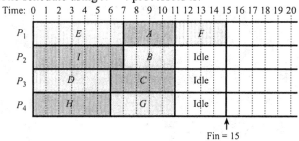

Fin = 15

(b)

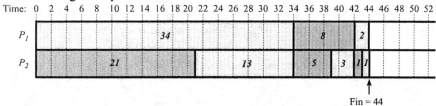

Fin = 12

(c) The relative error expressed as a percent is $\varepsilon = \dfrac{Fin - Opt}{Opt} = \dfrac{15 - 12}{12} = 25\%$.

57. (a) Since all tasks are independent, the critical path list is identical to a decreasing-time list. The schedule using $N = 2$ processors is:

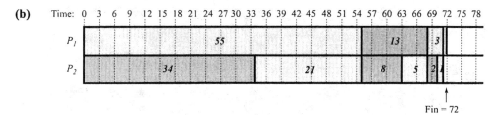

Fin = 44

(b)

Fin = 72

G. Miscellaneous

59. Every arc of the graph contributes 1 to the indegree sum and 1 to the outdegree sum.

61. (a) When $N = 7$, $\varepsilon = \dfrac{7-1}{3 \times 7} = \dfrac{6}{21} \approx 28.57\%$;

When $N = 8$, $\varepsilon = \dfrac{8-1}{3 \times 8} = \dfrac{7}{24} \approx 29.17\%$;

When $N = 9$, $\varepsilon = \dfrac{9-1}{3 \times 9} = \dfrac{8}{27} \approx 29.63\%$;

When $N = 10$, $\varepsilon = \dfrac{10-1}{3 \times 10} = \dfrac{9}{30} \approx 30\%$

(b) Since $M - 1 < M$, we have for every M that $\dfrac{M-1}{3M} \le \dfrac{M}{3M} = \dfrac{1}{3}$.

63. Refer to the project digraph given in Figure 8-15 of the text.

Time	Status of Processors		Priority List
	P_1	P_2	
20	Start: *IW*	Start: *PL*	~~AD~~ ~~AW~~ ~~AF~~ ~~IF~~ ~~AP~~ ~~IW~~ ID IP ⟨PL⟩ PU HU IC PD EU FW
24	Busy: *IW*	Idle	~~AD~~ ~~AW~~ ~~AF~~ ~~IF~~ ~~AP~~ ~~IW~~ ID IP ⟨PL⟩ PU HU IC PD EU FW
27	Start: *ID*	Start: *IP*	~~AD~~ ~~AW~~ ~~AF~~ ~~IF~~ ~~AP~~ ~~IW~~ ~~ID~~ ~~IP~~ ~~PL~~ PU HU IC PD EU FW
31	Busy: *ID*	Start: *HU*	~~AD~~ ~~AW~~ ~~AF~~ ~~IF~~ ~~AP~~ ~~IW~~ ~~ID~~ ~~IP~~ ~~PL~~ PU ~~HU~~ IC PD EU FW
32	Start: *PU*	Busy: *HU*	~~AD~~ ~~AW~~ ~~AF~~ ~~IF~~ ~~AP~~ ~~IW~~ ~~ID~~ ~~IP~~ ~~PL~~ ~~PU~~ ~~HU~~ IC PD EU FW
35	Start: *IC*	Start: *PD*	~~AD~~ ~~AW~~ ~~AF~~ ~~IF~~ ~~AP~~ ~~IW~~ ~~ID~~ ~~IP~~ ~~PL~~ ~~PU~~ ~~HU~~ ~~IC~~ ~~PD~~ ~~EU~~ ~~FW~~
36	Start: *EU*	Busy: *PD*	~~AD~~ ~~AW~~ ~~AF~~ ~~IF~~ ~~AP~~ ~~IW~~ ~~ID~~ ~~IP~~ ~~PL~~ ~~PU~~ ~~HU~~ ~~IC~~ ~~PD~~ ~~EU~~ ~~FW~~
38	Start: *FW*	Idle	~~AD~~ ~~AW~~ ~~AF~~ ~~IF~~ ~~AP~~ ~~IW~~ ~~ID~~ ~~IP~~ ~~PL~~ ~~PU~~ ~~HU~~ ~~IC~~ ~~PD~~ ~~EU~~ ~~FW~~
44	Idle	Idle	~~AD~~ ~~AW~~ ~~AF~~ ~~IF~~ ~~AP~~ ~~IW~~ ~~ID~~ ~~IP~~ ~~PL~~ ~~PU~~ ~~HU~~ ~~IC~~ ~~PD~~ ~~EU~~ ~~FW~~

The schedule is:

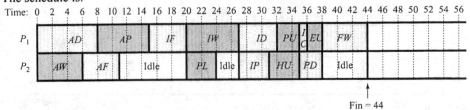

65. Refering to the project digraph with critical times in Figure 8-23 of the text, we see the critical path list is: *AP*[34], *AF*[32], *AW*[28], *IF*[27], *IW*[22], *AD*[18], *IP*[15], *PL*[11], *HU*[11], *ID*[10], *IC*[7], *FW*[6], *PU*[5], *PD*[3], *EU*[2].

Time	Status of Processors		Priority List
	P_1	P_2	
0	Start: *AP*	Start: *AF*	*AP AP AW* IF IW *AD* IP PL HU ID IC FW PU PD EU
5	Busy: *AP*	Start: *AW*	*AP AP AW* IF IW *AD* IP PL HU ID IC FW PU PD EU
7	Start: *IF*	Busy: *AW*	*AP AP AW IF* IW *AD* IP PL HU ID IC FW PU PD EU
11	Busy: *IF*	Start: *AD*	*AP AP AW IF IW AD* IP PL HU ID IC FW PU PD EU
12	Start: *IW*	Busy: *AD*	*AP AP AW IF IW AD* IP *PL* HU ID IC FW PU PD EU
19	Start: *IP*	Start: *PL*	*AP AP AW IF IW AD IP PL* HU *ID* IC FW PU PD EU
23	Start: *HU*	Start: *ID*	*AP AP AW IF IW AD IP PL* HU *ID* IC FW PU PD EU
27	Start: *IC*	Busy: *ID*	*AP AP AW IF IW AD IP PL HU ID IC* FW PU *PD* EU
28	Start: *FW*	Start: *PU*	*AP AP AW IF IW AD IP PL HU ID IC FW* PU *PD* EU
31	Busy: *FW*	Start: *PD*	*AP AP AW IF IW AD IP PL HU ID IC FW* PU *PD* *EU*
34	Start: *EU*	Idle	*AP AP AW IF IW AD IP PL HU ID IC FW PU PD* *EU*
36	Idle	Idle	*AP AP AW IF IW AD IP PL HU ID IC FW PU PD EU*

The schedule is:

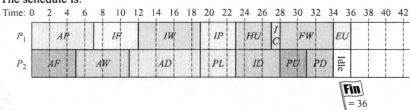

JOGGING

67. (a)

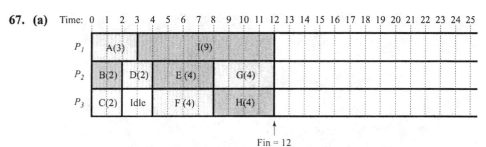

Fin = 12

(b)

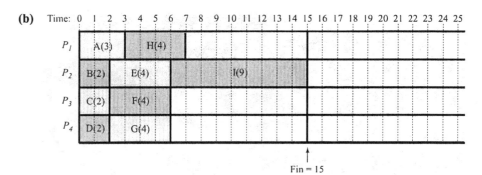

Fin = 15

(c) An extra processor was used and yet the finishing time of the project increased.

151

69. (a) This is the same question and answer as 67(a).

(b)

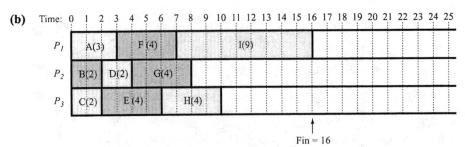

Fin = 16

(c) The project had fewer restrictions on the order of the assignments to the processors and yet the finishing time of the project increased.

71. (a) The finishing time of a project is always more than or equal to the number of hours of work to be done divided by the number of processors doing the work.

(b) The schedule is optimal with no idle time.

(c) The total idle time in the schedule.

Chapter 9

WALKING

A. Fibonacci Numbers

1. (a) $F_{10} = 55$

   (b) $F_{10} + 2 = 55 + 2 = 57$

   (c) $F_{10+2} = F_{12} = 144$

   (d) $F_{10} / 2 = 55 / 2 = 27.5$

   (e) $F_{10/2} = F_5 = 5$

3. (a) $F_1 + F_2 + F_3 + F_4 + F_5 = 1 + 1 + 2 + 3 + 5$
$$= 12$$

   (b) $F_{1+2+3+4+5} = F_{15} = 610$

   (c) $F_3 \times F_4 = 2 \times 3 = 6$

   (d) $F_{3 \times 4} = F_{12} = 144$

   (e) $F_{F_4} = F_3 = 2$

5. (a) $3F_N + 1$ represents one more than three times the Nth Fibonacci number.

   (b) $3F_{N+1}$ represents three times the Fibonacci number in position $(N + 1)$.

   (c) $F_{3N} + 1$ represents one more than the Fibonacci number in the $3N$th position.

   (d) F_{3N+1} represents the Fibonacci number in position $(3N + 1)$.

7. (a) $F_{38} = F_{37} + F_{36}$
$$= 24,157,817 + 14,930,352$$
$$= 39,088,169$$

   (b) $F_{35} = F_{37} - F_{36}$
$$= 24,157,817 - 14,930,352$$
$$= 9,227,465$$

9. (I) $F_{N+2} = F_{N+1} + F_N$, $N > 0$, is an equivalent way to express the fact that each term of the Fibonacci sequence is equal to the sum of the two preceding terms.

11. (a) $47 = 34 + 13$
Note that 34 is the largest Fibonacci number less than 47.

   (b) $48 = 34 + 13 + 1$
34 is the largest Fibonacci number less than 48. Then, note that 13 is the largest Fibonacci number less than the remaining $48 - 34 = 14$.

   (c) $207 = 144 + 55 + 8$
144 is the largest Fibonacci number less than 207. Next, 55 is the largest Fibonacci number less than the remaining $207 - 144 = 63$.

   (d) $210 = 144 + 55 + 8 + 3$
144 is the largest Fibonacci number less than 210. Also, 55 is the largest Fibonacci number less than the remaining $210 - 144 = 66$. Finally, 8 is the largest Fibonacci number less than the remaining $210 - 144 - 55 = 11$.

13. (a) Fifth equation in the sequence:
$1 + 2 + 5 + 13 + 34 + 89 = 144$

   (b) 22. In each case the Fibonacci number that appears on the right of the equality is that Fibonacci number immediately following the largest Fibonacci number appearing on the left.

   (c) $N+1$. $N+1$ is one bigger than N.

15. (a) Choosing, for example, the Fibonacci numbers $F_4 = 3$, $F_5 = 5$, $F_6 = 8$,

and $F_7 = 13$, the fact is verified by the equation
$F_7 + F_4 = 13 + 3 = 16 = 2(8) = 2F_6$.
This exercises asks you to verify four such examples.

(b) Denoting the first of these four Fibonacci numbers by F_N, a mathematical formula expressing this fact is $F_{N+3} + F_N = 2F_{N+2}$.

17. (a) $2F_6 - F_7 = 2 \times 8 - 13 = 3 = F_4$

(b) $2F_{N+2} - F_{N+3} = F_N$

B.   The Golden Ratio

19. (a) $21\left(\dfrac{1+\sqrt{5}}{2}\right) + 13 \approx 46.97871$

The order of operations is critical. First add 1 and $\sqrt{5}$. Next, divide by 2. Then multiply by 21. Finally, add 13.

(b) $\left(\dfrac{1+\sqrt{5}}{2}\right)^8 \approx 46.97871$

Order of operations is again critical. First add 1 and $\sqrt{5}$. Next, divide by 2. Finally, take the result to the power of 8.

(c) $\dfrac{\left(\frac{1+\sqrt{5}}{2}\right)^8 - \left(\frac{1-\sqrt{5}}{2}\right)^8}{\sqrt{5}} = 21.00000$

21. (a) $\dfrac{\phi^8}{\sqrt{5}} = \dfrac{\left[\frac{(1+\sqrt{5})}{2}\right]^8}{\sqrt{5}} \approx 21$

Order of operations is critical. First add 1 and $\sqrt{5}$. Next, divide by 2. Next, take the result to the power of 8. Finally, divide the result by $\sqrt{5}$.

(b) $\dfrac{\phi^9}{\sqrt{5}} = \dfrac{\left[\frac{(1+\sqrt{5})}{2}\right]^9}{\sqrt{5}} \approx 34$

(c) Guess the 7th Fibonacci number, 13.

23. (a) $\phi^9 = F_9\phi + F_8$
$= 34\phi + 21$
so $a = 34$ and $b = 21$.

(b) $\phi^9 = F_9\phi + F_8$
$= 34\phi + 21$
$= 34\left(\dfrac{1+\sqrt{5}}{2}\right) + 21$
$= 17 + 17\sqrt{5} + 21$
$= 38 + 17\sqrt{5}$
So, $a = 17$ and $b = 38$.

25. $F_{500} \approx \phi \times F_{499}$
$\approx 1.618 \times 8.617 \times 10^{103}$
$\approx 1.394 \times 10^{104}$

27. (a) $A_7 = 2A_{7-1} + A_{7-2}$
$= 2A_6 + A_5$
$= 2(70) + 29$
$= 169$

(b) $\dfrac{A_7}{A_6} = \dfrac{169}{70}$
≈ 2.41429

(c) $\dfrac{A_{11}}{A_{10}} = \dfrac{5,741}{2,378}$
≈ 2.41421

(d) 2.41421 (Though not obvious, the ratio actually approaches $1 + \sqrt{2}$ as the value of N increases.)

C.   **Fibonacci Numbers and Quadratic Equations**

29. (a) $x^2 = 2x + 1$

$$x^2 - 2x - 1 = 0$$

$$x = \frac{-(-2) \pm \sqrt{(-2)^2 - 4(1)(-1)}}{2(1)}$$

$$= \frac{2 \pm \sqrt{8}}{2}$$

$$= \frac{2 \pm 2\sqrt{2}}{2}$$

$$= 1 \pm \sqrt{2}$$

$$x = 1 + \sqrt{2} \quad \text{or} \quad x = 1 - \sqrt{2}$$

$$\approx 2.41421 \qquad\qquad \approx -0.41421$$

(b) $3x^2 = 5x + 8$

$$3x^2 - 5x - 8 = 0$$

$$x = \frac{-(-5) \pm \sqrt{(-5)^2 - 4(3)(-8)}}{2(3)}$$

$$= \frac{5 \pm \sqrt{121}}{6}$$

$$= \frac{5 \pm 11}{6}$$

$$x = \frac{5 + 11}{6} \quad \text{or} \quad x = \frac{5 - 11}{6}$$

$$= \frac{16}{6} \qquad\qquad = \frac{-6}{6}$$

$$= \frac{8}{3} \qquad\qquad = -1$$

$$\approx 2.66667$$

31. (a) Try $x = 1$:

$$55(1)^2 = 34(1) + 21$$

$$55 = 55$$

$$x = 1 \text{ is one solution.}$$

(b) Rewrite the equation:

$$55x^2 - 34x - 21 = 0$$

Then

$$1 + x = \frac{34}{55}$$

$$x = \frac{34}{55} - 1$$

$$= -\frac{21}{55}$$

$$\approx -0.38182$$

33. (a) Putting $x = 1$ in the equation gives $F_N = F_{N-1} + F_{N-2}$ which is a defining equation for the Fibonacci numbers.

(b) Rewrite the equation:

$$F_N x^2 - F_{N-1} x - F_{N-2} = 0$$

Then, using the hint in Exercise 31(b),

$$1 + x = \frac{F_{N-1}}{F_N}$$

$$x = \left(\frac{F_{N-1}}{F_N} \right) - 1$$

D.   **Similarity**

35. (a) Since R and R' are similar, each side length of R' is 3 times longer than the corresponding side in R. So, the perimeter of R' will be 3 times greater than the perimeter of R. This means that the perimeter of R' is $3 \times 41.5 = 124.5$ inches.

(b) Since each side of R' is 3 times longer than the corresponding side in R, the area of R' will be $3^2 = 9$ times larger than the area of R. That is, the area of R' is $9 \times 105 = 945$ square inches.

37. (a) The ratio of the perimeters must be the same as the ratio of corresponding side lengths.

$$\frac{P}{13 \text{ in.}} = \frac{60 \text{ m}}{5 \text{ in.}}$$

$$P = \frac{60 \text{ m}}{5 \text{ in.}} \times 13 \text{ in.}$$

$$= 156 \text{ m}$$

(b) In this problem we use the fact that the ratio of the areas of two similar triangles (or any two similar polygons) is the same as the ratio of their side lengths squared. Hence,

$$\frac{A}{20 \text{ sq. in.}} = \left(\frac{60 \text{ m}}{5 \text{ in.}}\right)^2$$

$$A = \left(\frac{60 \text{ m}}{5 \text{ in.}}\right)^2 \times (20 \text{ sq. in.})$$

$$= 2880 \text{ sq. m}$$

39. There are two possible cases that need to be considered. First, we solve

$$\frac{3}{x} = \frac{5}{8-x}$$

$$24 - 3x = 5x$$

$$24 = 8x$$

$$3 = x$$

But, it could also be the case that the side of length 3 does not correspond to the side of length 5, but rather the side of length $8 - x$. In this case, we solve

$$\frac{3}{x} = \frac{8-x}{5}$$

$$15 = 8x - x^2$$

$$x^2 - 8x + 15 = 0$$

$$(x-5)(x-3) = 0$$

$$x = 3, 5$$

It follows that R and R' are similar if $x = 3$ or if $x = 5$ (as can easily be checked).

E.   Gnomons

41.
$$\frac{3}{9} = \frac{9}{c+3}$$

$$3(c+3) = 81$$

$$3c + 9 = 81$$

$$3c = 72$$

$$c = 24$$

43.
$$\frac{8}{12} = \frac{3+8+1}{2+12+x}$$

$$\frac{8}{12} = \frac{12}{14+x}$$

$$8(14 + x) = 144$$

$$112 + 8x = 144$$

$$8x = 32$$

$$x = 4$$

45.

$$\frac{10}{20} = \frac{20}{10+x}$$

$$10(10 + x) = 400$$

$$100 + 10x = 400$$

$$10x = 300$$

$$x = 30$$

Rectangle B is 20 by 30.

47. (a) In the figure, the measure of angle CAD must be $180° - 108° = 72°$. Since triangle BDC must be similar to triangle BCA (in order for triangle ACD to be a gnomon), it must be that the measure of angle BDC must be $36°$. Using the fact that the sum of the measures of the angles in any triangle is $180°$, it follows that the measure of angle ACD is also

$36°$.

C

$36°$

ϕ x

$36°$ $108°$

B   1   A   y   D

(b) Since triangle DBC is isosceles, $x = \phi$.
Note that triangle ACD is also isosceles.
Hence, $y = x = \phi$ as well.

49. $\dfrac{3}{4} = \dfrac{9}{x}$ and $\dfrac{3}{5} = \dfrac{9}{5+y}$

$3x = 36$ and $3(5+y) = 45$

$x = 12$ and $y = 10$

JOGGING

51. $A_N = 5F_N$ (Each term in this sequence is 5 times more than the corresponding term in the Fibonacci sequence.)

53. (a) $T_1 = 7F_2 + 4F_1 = 7 \cdot 1 + 4 \cdot 1 = 11$
$T_2 = 7F_3 + 4F_2 = 7 \cdot 2 + 4 \cdot 1 = 18$
$T_3 = 7F_4 + 4F_3 = 7 \cdot 3 + 4 \cdot 2 = 29$
$T_4 = 7F_5 + 4F_4 = 7 \cdot 5 + 4 \cdot 3 = 47$
$T_5 = 7F_6 + 4F_5 = 7 \cdot 8 + 4 \cdot 5 = 76$
$T_6 = 7F_7 + 4F_6 = 7 \cdot 13 + 4 \cdot 8 = 123$
$T_7 = 7F_8 + 4F_7 = 7 \cdot 21 + 4 \cdot 13 = 199$
$T_8 = 7F_9 + 4F_8 = 7 \cdot 34 + 4 \cdot 21 = 322$

(b) $T_N = 7F_{N+1} + 4F_N$
$= 7(F_N + F_{N-1}) + 4(F_{N-1} + F_{N-2})$
$= (7F_N + 4F_{N-1}) + (7F_{N-1} + 4F_{N-2})$
$= T_{N-1} + T_{N-2}$

(c) $T_1 = 11$, $T_2 = 18$,
$T_N = T_{N-1} + T_{N-2}$

55. $\dfrac{1+\sqrt{5}}{2} + \dfrac{1-\sqrt{5}}{2} = \dfrac{1+1}{2} = 1$
Since the first term is positive and the second term is negative and the sum is a whole number, it follows that the decimal parts are the same.

57. We use the fact that $\phi^N = F_N\phi + F_{N-1}$ to compute the ratio l/s. Since
$\dfrac{l}{s} = \dfrac{144\phi + 89}{89\phi + 55} = \dfrac{\phi^{12}}{\phi^{11}} = \phi$, it follows that R is a golden rectangle.

59. If the rectangle is a gnomon to itself, then the rectangle in the figure below must be similar to the original l by s rectangle.

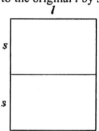

This means that $\dfrac{2s}{l} = \dfrac{l}{s}$. Letting $x = \dfrac{l}{s}$, we

solve $\dfrac{2}{x} = x$. The only positive solution to

this equation is $x = \dfrac{l}{s} = \sqrt{2}$. So, the only

rectangles that are gnomons to themselves (homognomic) are those having this ratio.

61. Since the area of the white triangle is 6, the area of the shaded figure must be 48, which makes the area of the new larger similar triangle 54. Since the ratio of the areas of similar triangles is the square of the ratio of the sides, we have

$\dfrac{3+x}{3} = \sqrt{\dfrac{54}{6}} = 3$ and $\dfrac{y}{4} = \sqrt{\dfrac{54}{6}} = 3$ and

$\dfrac{5+z}{5} = \sqrt{\dfrac{54}{6}} = 3$ so $x = 6$, $y = 12$, $z = 10$.

63. Since the area of the white rectangle is 60 and the area of the shaded figure is 75, the area of the new larger similar rectangle is 135. Since the ratio of the area of similar rectangles is the square of the ratio of the sides, we have

$$\frac{6+x}{6} = \sqrt{\frac{135}{60}} = \sqrt{\frac{9}{4}} = \frac{3}{2} \text{ and } \frac{10+y}{10} = \frac{3}{2} \text{ so}$$
$x = 3, y = 5.$

65. **(a)** Since we are given that $AB = BC = 1$, we know that $\angle BAC = 72°$ and so $\angle BAD = 180° - 72° = 108°$. This makes $\angle ABD = 180° - 108° - 36° = 36°$ and so triangle ABD is isosceles with $AD = AB = 1$. Therefore $AC = x - 1$. Using these facts and the similarity of triangle ABC and triangle BCD we have $\frac{x}{1} = \frac{1}{x-1}$ or,

$x^2 = x + 1$ for which we know the solution is $x = \phi$.

(b) 36°, 36°, 108°

(c) $\frac{\text{longer side}}{\text{shorter side}} = \frac{x}{1} = x = \phi$

67. **(a)** The regular decagon can be split into ten $72° - 72° - 36°$ triangles as shown in the figure. In each of these triangles, the side opposite the 72° angle is the same length as the radius of the circle, namely r. In Exercise 65, it was shown that the length of the side opposite the

36° angle is $\frac{r}{\phi}$. Hence, the perimeter

of the regular decagon is $10 \times \frac{1}{\phi} = \frac{10}{\phi}$.

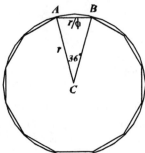

(b) Using part (a), the perimeter is found to be $\frac{10r}{\phi}$.

Chapter 10

WALKING

A. Linear Growth and Arithmetic Sequences

1. (a) $P_1 = P_0 + 125 = 80 + 125 = 205$
$P_2 = P_1 + 125 = 205 + 125 = 330$
$P_3 = P_2 + 125 = 330 + 125 = 455$

(b) $P_{100} = 80 + 125 \times 100 = 12,580$

(c) $P_N = 80 + 125N$

3. (a) $P_{30} = 74 + 5 \times 30 = 225$

(b) $1000 \le 75 + 5N$
$925 \le 5N$
$N \ge 185$

(c) The population will reach (and surpass) 1002 in the generation after it reaches 1000. Thus, it will take 186 generations for the population to reach 1002.

5. (a) $38 = 8 + 10 \times d$
$30 = 10d$
$d = 3$

(b) $P_{50} = 8 + 50 \times 3 = 158$

(c) $P_N = 8 + 3N$

7. (a) $d = A_2 - A_1 = -4 - 11 = -15$
$A_3 = A_2 + (-15) = -4 - 15 = -19$

(b) $A_1 = A_0 + (-15)$
$11 = A_0 - 15$
$A_0 = 26$

(c) None. The sequence is decreasing and starts at 26.

9. (a) $P_N = P_{N-1} + 5; P_0 = 3$

(b) $P_N = 3 + 5N$

(c) $P_{300} = 3 + 5 \times 300 = 1503$

11. (a) $2 + 5 \times 99 = 497$

(b) $A_0 = 2, \ d = 5$
The 100th term is
$A_{99} = 2 + 5 \times 99 = 497$.
This is the sum of 100 terms in an arithmetic sequence:
$2 + 7 + 12 + \ldots + 497 = \dfrac{(2 + 497) \times 100}{2}$
$= 24,950$

13. (a) $d = 3$
If 309 is the Nth term of the sequence,
$309 = 12 + 3(N - 1)$
$309 = 12 + 3N - 3$
$300 = 3N$
$N = 100$
So, 309 is the 100th term of the sequence.

(b) $12 + 15 + 18 + \ldots + 309 = \dfrac{(12 + 309) \times 100}{2}$
$= 16,050$

15. (a) $P_{999} = 23 + 7 \times 999 = 7016$
$P_0 + P_1 + P_2 + \ldots + P_{999}$
$= \dfrac{(23 + 7016) \times 1000}{2}$
$= 3,519,500$

(b) $P_{99} = 23 + 7 \times 99 = 716$
$P_0 + P_1 + P_2 + \ldots + P_{99} = \dfrac{(23 + 716) \times 100}{2}$
$= 36,950$

$$P_{100} + P_{101} + \ldots + P_{999}$$
$$= (P_0 + P_1 + P_2 + \ldots + P_{999})$$
$$\quad - (P_0 + P_1 + P_2 + \ldots + P_{99})$$
$$= 3,519,500 - 36,950$$
$$= 3,482,550$$

17. (a) $P_{38} = 137 + 2 \times 38 = 213$

(b) $P_N = 137 + 2N$

(c) $137 \times \$1 \times 52 = \$7124$

(d) When just counting the newly installed lights, $P_0 = 0$ and $P_{51} = 2 \times 51 = 102$.
(The lights installed in the 52nd week aren't in operation during the 52-week period.)
$$0 + 2 + 4 + \ldots + 102 = \frac{(0 + 102) \times 52}{2}$$
$$= 2652$$
The cost is \$2,652.

B. Exponential Growth and Geometric Sequences

19. (a) $P_1 = 11 \times 1.25 = 13.75$

(b) $P_9 = 11 \times (1.25)^9 \approx 81.956$

(c) $P_N = 11 \times (1.25)^N$

21. (a) $P_1 = 4 \times P_0 = 4 \times 5 = 20$
$$P_2 = 4 \times P_1 = 4 \times 20 = 80$$
$$P_3 = 4 \times P_2 = 4 \times 80 = 320$$

(b) $P_N = 5 \times 4^N$

(c) 9 generations
We solve $P_N = 5 \times 4^N \geq 1,000,000$ for N. By trial and error, we find that
$$5 \times 4^8 = 327,680 \text{ and}$$
$$5 \times 4^9 = 1,310,720.$$

23. (a) $P_N = 1.50 P_{N-1}$ where $P_0 = 200$
We multiply by 1.50 each year to account for an *increase* by 50%.

(b) $P_N = 200 \times 1.5^N$

(c) $P_{10} = 200 \times 1.5^{10} \approx 11,533$
A good estimate would be to say that about 11,500 crimes will be committed in 2010.

25. (a) $P_{100} = 3 \times 2^{100}$

(b) $P_N = 3 \times 2^N$

(c) $a = 3, r = 2, N = 101$.
$$P_0 + P_1 + \ldots + P_{100}$$
$$= 1 + 1 \times 2 + 1 \times 2^2 + \ldots + 1 \times 2^{100}$$
$$= \frac{3 \times (2^{101} - 1)}{2 - 1}$$
$$= 3 \times (2^{101} - 1)$$

(d) $P_0 + P_1 + \ldots + P_{49} = \dfrac{3 \times (2^{50} - 1)}{2 - 1}$
$$= 3 \times (2^{50} - 1)$$
$$P_{50} + P_{51} + \ldots + P_{100}$$
$$= (P_0 + P_1 + \ldots + P_{100}) - (P_0 + P_1 + \ldots + P_{49})$$
$$= 3 \times (2^{101} - 1) - 3 \times (2^{50} - 1)$$
$$= 3 \times 2^{101} - 3 \times 2^{50}$$
$$= 3 \times 2^{50} \times (2^{51} - 1)$$

C. Financial Applications

27. (a) Case 1: Coupon before 30% discount
Price after coupon:
$100 - 0.15 \times (\$100) = \$85$
Final price:
$85 - 0.30 \times (\$85) = \$59.50$
Amount of discount:
$100 - \$59.50 = \$40.50$
Case 2: Coupon after 30% discount
Price after discount:
$100 - 0.30 \times (\$100) = \$70$

Final price:
$70 - 0.15 \times \$70 = \$59.50$
Amount of discount:
$100 - \$59.50 = \$40.50$

(b) $\dfrac{\$40.50}{\$100} \times 100\% = 40.5\%$

(c) Case 1: Coupon before 30% discount
Price after coupon:
$P - 0.15P = 0.85P$
Final price:
$0.85P - 0.30(0.85P) = 0.85P - 0.255P$
$\qquad\qquad\qquad\qquad = 0.595P$
Amount of discount:
$P - 0.595P = 0.405P$
Percent of discount is 40.5%.

Case 2: Coupon after 30% discount
Price after discount:
$P - 0.30P = 0.70P$
Final price:
$0.70P - 0.15(0.70P) = 0.70P - 0.105P$
$\qquad\qquad\qquad\qquad = 0.595P$
Amount of discount:
$P - 0.595P = 0.405P$
Percent of discount is 40.5%.
The percentage discount does not depend on the original price.

29. $P_4 = \$3250 \times (1.09)^4 = \$4587.64$

31. The initial deposit of $3420 grows at a $6\frac{5}{8}\%$ annual interest rate for 3 years. At the end of three years (on Jan. 1, 2009), the amount in the account is $\$3420 \times (1.06625)^3 = \$4145.75$. This amount then grows at a $5\frac{3}{4}\%$ annual interest rate for 4 years. At the end of that time (on Jan. 1, 2013), the amount in the account is $\$4145.75 \times (1.0575)^4 = \$5184.71$.

33. The balance on December 31, 2004 is
$P_2 = \$3420 \times (1.06625)^2 = \$3888.16$.

The balance on January 1, 2005 is
$\$3888.16 - \$1500 = \$2388.16$.
The balance on December 31, 2005 is
$\$2388.16 \times (1.06625)^1 = \$2546.38$.
The balance on January 1, 2006 is
$\$2546.38 - \$1000 = \$1546.38$.
The balance on January 1, 2009 is
$\$1546.38 \times (1.06625)^3 = \$1874.53$.

35. (a) The periodic interest rate is given by
$$p = \frac{0.12}{12} = 0.01 \ \text{(i.e. 1%)}.$$

(b) The amount in the account after 5 years (60 months) is
$$P_{60} = \$5000 \times (1.01)^{60} = \$9083.48$$

(c) Each $1 invested grows to
$$\$\left(1 + \frac{0.12}{12}\right)^{12} \approx \$1.126825$$
So, the annual yield is approximately 12.6825%.

37. Great Bulldog Bank: 6%;
First Northern Bank:
Each $1 invested grows to
$$\$\left(1 + \frac{0.0575}{12}\right)^{12} \approx \$1.05904$$
The annual yield is thus 5.904%;
Bank of Wonderland:
Each $1 invested grows to
$$\$\left(1 + \frac{0.055}{365}\right)^{365} \approx \$1.05654$$
The annual yield is thus 5.654%.

39. (a) $\$1000 \times (1.1)^{25} = \$10,834.71$

(b) $\$1000 \times (1.102)^{25} = \$11,338.09$

(c) $\$1000 \times (1.0996)^{25} = \$10,736.64$

41. The periodic interest rate is
$$p = \frac{0.06}{12} = 0.005.$$

The Jan. 1 deposit will grow to $100(1.005)^{11}$.
The Feb. 1 deposit will grow
to $100(1.005)^{10}$. The Mar. 1 deposit will
grow to $100(1.005)^{9}$, and so on. The total
amount in the account will be
$100(1.005)+100(1.005)^2+...+100(1.005)^{11}$.
This is the sum of $N = 11$ terms of a
geometric sequence having first term
$a=100(1.005)$ and common ratio $r=1.005$.
The sum is
$$\frac{100(1.005)[(1.005)^{11}-1]}{1.005-1} \approx \$1133.56.$$

43. (a) The amount after 5 years is
    $P_5 = \$10,000$. The periodic interest
    rate is $p = 0.10$. The number of years is
    $N = 5$. Thus, we solve for the initial
    investment P_0.
    $$10,000 = P_0 \times (1.10)^5$$
    $$P_0 \approx 6209.21$$

    (b) $P_5 = \$10,000,\ p = \frac{0.10}{4} = 0.025,$
    $N = 20$
    $$10,000 = P_0 \times (1.025)^{20}$$
    $$P_0 \approx \$6102.71$$

    (c) $P_5 = \$10,000, p = \frac{0.10}{12},\ N = 60$
    $$10,000 = P_0 \times \left(1+\frac{0.10}{12}\right)^{60}$$
    $$P_0 \approx \$6077.89$$

D.   Logistic Growth Model

45. (a) $p_1 = 0.8 \times (1-0.3) \times 0.3 = 0.1680$

    (b) $p_2 = 0.8 \times (1-0.168) \times 0.168 \approx 0.1118$

    (c) $p_3 = 0.8 \times (1-0.11182) \times 0.11182$
    ≈ 0.07945
    Thus, approximately 7.945% of the

habitat's carrying capacity is taken up
by the third generation.

47. (a) Using the formula $p_{N+1} = r(1-p_N)p_N$
    and a calculator with a memory register or
    a spreadsheet, we get
    $p_1 = 0.1680, p_2 \approx 0.1118, p_3 \approx 0.0795,$
    $p_4 \approx 0.0585, p_5 \approx 0.0441, p_6 \approx 0.0337,$
    $p_7 \approx 0.0261, p_8 \approx 0.0203, p_9 \approx 0.0159,$
    $p_{10} \approx 0.0125.$

    (b) Since $p_N \to 0$ this logistic growth
    model predicts extinction for this
    population.

49. (a) Using the formula $p_{N+1} = r(1-p_N)p_N$
    and a calculator with a memory register or
    a spreadsheet, we get
    $p_1 = 0.4320, p_2 \approx 0.4417, p_3 \approx 0.4439,$
    $p_4 \approx 0.4443, p_5 \approx 0.4444, p_6 \approx 0.4444,$
    $p_7 \approx 0.4444, p_8 \approx 0.4444, p_9 \approx 0.4444,$
    $p_{10} \approx 0.4444.$

    (b) The population becomes stable at
    $\frac{4}{9} \approx 44.44\%$ of the habitat's carrying
    capacity.

51. (a) $p_1 = 0.3570, p_2 \approx 0.6427, p_3 \approx 0.6429,$
    $p_4 \approx 0.6428, p_5 \approx 0.6429, p_6 \approx 0.6428,$
    $p_7 \approx 0.6429, p_8 \approx 0.6428, p_9 \approx 0.6429,$
    $p_{10} \approx 0.6428$

    (b) The population becomes stable at
    $\frac{9}{14} \approx 64.29\%$ of the habitat's carrying
    capacity.

53. (a) $p_1 = 0.5200, p_2 = 0.8112, p_3 \approx 0.4978,$
    $p_4 \approx 0.8125, p_5 \approx 0.4952, p_6 \approx 0.8124,$
    $p_7 \approx 0.4953, p_8 \approx 0.8124, p_9 \approx 0.4953,$
    $p_{10} \approx 0.8124$

(b) The population settles into a two-period cycle alternating between a high-population period at approximately 81.24% and a low-population period at approximately 49.53% of the habitat's carrying capacity.

E.   Miscellaneous

55. (a) Exponential ($r = 2$)

(b) Linear ($d = 2$)

(c) Logistic

(d) Exponential ($r = \dfrac{1}{3}$)

(e) Logistic

(f) Linear ($d = -0.15$)

(g) Linear ($d = 0$), Exponential ($r = 1$), and/or Logistic (they all apply!)

57. (a) $P_2 = 10 + 2 \times 6 = 22$
$P_3 = 22 + 2 \times 10 = 42$

(b) The first two numbers in this sequence are even. Thereafter, the sum and product of two even numbers is also even.

JOGGING

59. Given the initial deposit of $P_0 = \$500$, we solve $P_2 = P_0 (1+r)^2$ for the annual yield r. But,

$561.80 = 500(1+r)^2$
$561.80 = 500r^2 + 1000r + 500$
$0 = 500r^2 + 1000r - 61.80$

and the quadratic equation gives

$$r = \frac{-1000 \pm \sqrt{1000^2 - 4(500)(-61.8)}}{2(500)}$$
$$= \frac{-1000 \pm \sqrt{1,123,600}}{1000}$$
$$= \frac{-1000 \pm 1060}{1000}$$
$$= \frac{60}{1000}, \frac{-2060}{1000}$$
$$= 0.06, -2.06$$

Taking the positive value of r gives an annual yield of 6%.

61. (a) A 6-8-10 triangle.
Suppose, more generally, that a is the length of the shortest side of such a triangle and the other sides are of length $a+d$ and $a+2d$. By the Pythagorean theorem,

$$a^2 + (a+d)^2 = (a+2d)^2$$
$$a^2 + a^2 + 2ad + d^2 = a^2 + 4ad + 4d^2$$
$$a^2 - 2ad - 3d^2 = 0$$
$$(a-3d)(a+d) = 0$$

So, $a = 3d$ or $a = -d$. Since $a>0$, the only right triangles for which the sides are in an arithmetic sequence are 3*d*-4*d*-5*d* triangles.

(b) A $1-\sqrt{\Phi}-\Phi$ triangle.
The shortest side of such a triangle is 1 unit (by an appropriate choice of units). By the Pythagorean theorem, when the sides are in a geometric sequence with common ratio k, we must have $1^2 + k^2 = k^4$. Let $x = k^2$. Then $1 + x = x^2$. So, $x = \dfrac{1 \pm \sqrt{5}}{2}$. Since $x > 0$, we have $x = k^2 = \dfrac{1+\sqrt{5}}{2}$ (the golden ratio Φ). So $k = \sqrt{\Phi}$ and $k^2 = \Phi$. So, right triangles for which the sides are in

a geometric sequence are $1 - \sqrt{\Phi} - \Phi$ triangles.

63. If T was the starting tuition, then the tuition at the end of one year is 110% of T. That is, $(1.10)T$. The tuition at the end of two years is 115% of what it was after one year. That is, after two years, the tuition was $(1.15)(1.10)T$. Likewise, the tuition at the end of the three years was $(1.10)(1.15)(1.10)T = 1.3915T$. The total percentage increase was 39.15%.

65. We sum $\underbrace{0.01 + 0.02 + 0.04 + 0.08 + ...}_{\text{30 terms}}$

This is a sum of terms in a geometric sequence having $a = 0.01$, $r = 2$, and $N = 30$. We compute

$0.01 + 0.01 \times 2 + 0.01 \times 2^2 + ... + 0.01 \times 2^{29}$

$= \dfrac{0.01 \times (2^{30} - 1)}{2 - 1}$

$= \$10,737,418.23$

67. $r = \dfrac{1 \pm \sqrt{5}}{2}$

If a is a term of the geometric sequence with common ratio r, then ar and ar^2 are the following two terms. From the recursive rule we get $ar^2 = ar + a$. Assuming $a \neq 0$, this is equivalent to $r^2 = r + 1$, which gives us $r = \dfrac{1 \pm \sqrt{5}}{2}$.

69. $\left(a + ar + ar^2 + ... + ar^{N-1}\right)(r - 1)$

$= \quad ar + ar^2 + ar^3 + ... + ar^{N-1} + ar^N$

$\quad - a - ar - ar^2 - ar^3 - ... - ar^{N-1}$

$= -a + ar^N$

$= a(r^N - 1)$

So, after dividing both sides by $(r-1)$, we

arrive at

$a + ar + ar^2 + ... + ar^{N-1} = \dfrac{a(r^N - 1)}{r - 1}$.

71. One example is given by $P_N = 8 \times \left(\dfrac{1}{2}\right)^N$.

Here $r = \dfrac{1}{2}$ and $P_0 = 8$, $P_1 = 4$, $P_2 = 2$,

$P_3 = 1$, $P_4 = \dfrac{1}{2}, P_5 = \dfrac{1}{4}...$

73. No. This would require $p_0 = p_1 = 0.8(1 - p_0)p_0$ and $p_0 = 0$ or $1 = 0.8(1 - p_0)$. So, $p_0 = 0$ or $p_0 = -0.25$, neither of which are possible.

75. $p_0 = \dfrac{5000}{20,000} = 0.25$

$p_1 = 3.0 \times (1 - 0.25) \times 0.25 = 0.5625$

$p_2 \approx 0.7383$

$p_3 \approx 0.5797$

$p_4 \approx 0.7310$

$p_4 \times 20,000 \approx 14,619$ snails

Mini-Excursion 3

WALKING

A.   The Geometric Sum Formula

  1. In this sum, we have $a = 10$, $r = 1.05$, and $N = 36$. Thus,

$$\$10 + \$10(1.05) + \$10(1.05)^2 + \ldots + \$10(1.05)^{35} = \$10\frac{1.05^{36} - 1}{1.05 - 1}$$
$$= \$958.36$$

  3. In this sum, we have $a = 10(1.05)^{-35}$, $r = 1.05$, and $N = 36$. Thus,

$$\$10 + \$10(1.05)^{-1} + \$10(1.05)^{-2} + \ldots + \$10(1.05)^{-35} = \$10(1.05)^{-35}\frac{1.05^{36} - 1}{1.05 - 1}$$
$$= \$10\frac{1.05 - (1.05)^{-35}}{0.05}$$
$$= \$173.74$$

We could also take $a = 10$, $r = (1.05)^{-1}$, and $N = 36$ giving

$$\$10 + \$10(1.05)^{-1} + \$10(1.05)^{-2} + \ldots + \$10(1.05)^{-35} = \$10\frac{(1.05)^{-36} - 1}{(1.05)^{-1} - 1}$$
$$= \$173.74$$

  5. In this sum, we can take $a = 10(1.05)^{-36}$, $r = 1.05$, and $N = 36$. Thus,

$$\$10(1.05)^{-1} + \$10(1.05)^{-2} + \ldots + \$10(1.05)^{-35} + \$10(1.05)^{-36} = \$10(1.05)^{-36}\frac{1.05^{36} - 1}{1.05 - 1}$$
$$= \$10\frac{1 - (1.05)^{-36}}{0.05}$$
$$= \$165.47$$

  7. (a) In this sum, we take $a = 399(1.0075)^{-72}$, $r = 1.0075$, and $N = 72$. Then, by the geometric sum

formula, $\$399(1.0075)^{-1} + \$399(1.0075)^{-2} + \ldots + \$399(1.0075)^{-72} = \$399(1.0075)^{-72}\dfrac{(1.0075)^{72} - 1}{1.0075 - 1}$.

  (b) To obtain the right hand side of this equation from the left hand side, distribute $(1.0075)^{-72}$ and note that $(1.0075)^{-72}(1.0075)^{72} = 1$.

B.   Annuities, Investments and Loans

9. Markus' first $2,000 deposit grows for 39 years to a future value of $\$2000(1.075)^{39}$. His second $2,000 deposit grows for 38 years to a future value of $\$2000(1.075)^{38}$. His third $2,000 deposit grows for 37 years to a future value of $\$2000(1.075)^{37}$. And so on. His second to last deposit of $2,000 grows for one year to a future value of $\$2000(1.075)^{1}$. Markus' last deposit is $2,000 and does not grow.

The future value of this deferred annuity is the sum of the future values of all the deposits. In this case, that is $\$2000(1.075)^{39} + \$2000(1.075)^{38} + \ldots + \$2000(1.075)^{1} + \$2000$. Writing this sum in the reverse order allows us to recognize that $a = PMT = \$2000$, $r = 1.075$, and $N = 40$. So,

$$FV = a\frac{r^{N}-1}{r-1} = \$2000\frac{(1.075)^{40}-1}{1.075-1} \approx \$454,513.04.$$

11. The periodic interest rate in this problem is $p = 0.06/12 = 0.005$ so that $r = 1 + p = 1.005$. The number of periods N that Donald will make a payment is $N = 12 \times 35 = 420$. We solve the formula for the future value of a fixed deferred annuity $FV = a\frac{r^{N}-1}{r-1}$ where $FV = \$1,000,000$, $r = 1.005$, and $N = 420$ for $a = PMT$. This gives $a = PMT = FV\frac{r-1}{r^{N}-1} = \$1,000,000\frac{1.005-1}{(1.005)^{420}-1} \approx \$701.90$. Donald should sock away a little more than $700 each month.

13. Since Freddy only made a single deposit, we can use the compound interest formula to give us the future value of the account. Using $i = 0.07$ and $N = 15$, we solve $\$1172.59 = P_0(1.07)^{15}$ for Freddy's initial deposit P_0. But then $P_0 = \frac{\$1172.59}{(1.07)^{15}} = \$425.00$.

15. We find the present value of a fixed immediate annuity. In this case, the payment is $PMT = \$100,000,000$, $N = 10$ years, and $p = 0.15$. Hence

$$PV = \$PMT\frac{1-(1+p)^{-N}}{p} = \$100,000,000\frac{1-(1.15)^{-10}}{0.15} = \$501,876,863.$$

17. We find the present value of a fixed immediate annuity. Here the payment is $PMT = \$50$, $N = 52 \times 60 = 3120$ weeks, and $p = 0.055/52$. Hence

$$PV = \$PMT\frac{1-(1+p)^{-N}}{p} = \$50\frac{1-\left(1+\frac{0.055}{52}\right)^{-3120}}{\frac{0.055}{52}} = \$45,526.11.$$

19. (a) The amortization formula can be used the find the payment for this fixed immediate annuity. The present value of the loan is $160,000, $N = 12 \times 30 = 360$ months, and the periodic rate

is $p = 0.0575/12$. Hence

$$PMT = \$PV \frac{p}{1-(1+p)^{-N}} = \$160,000 \frac{\dfrac{0.0575}{12}}{1-\left(1+\dfrac{0.0575}{12}\right)^{-360}} = \$933.72.$$

(b) The Simpsons will make 360 payments of \$933.72 for a total of \$336,139.20. This means they will pay \$336,139.20 - \$160,000 = \$176,139.20 in interest.

21. We find the present value of a fixed immediate annuity and then add \$25,000. In this case, the payment is $PMT = \$950$, $N = 12 \times 20 = 240$ months, and $p = 0.055/12$. So,

$$PV = \$PMT \frac{1-(1+p)^{-N}}{p} = \$950 \frac{1-\left(1+\dfrac{0.055}{12}\right)^{-240}}{\dfrac{0.055}{12}} = \$138,104.01.$$ Adding the \$25,000 down

payment, we find the selling price of the house that Ken just bought to be approximately \$163,104.

❧

Chapter 11

WALKING

A.   Reflections

1. (a) *C.* The image point of *P* under the reflection with axis l_1 is found by drawing a line through *P* perpendicular to l_1 and finding the point on this line on the opposite side of l_1 which is the same distance from l_1 as the point *P*. This point is *C*.

(b) *F*

(c) *E*

(d) *B*

3. (a)
(b)
(c)

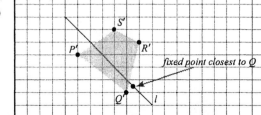

5. (a)
(b)
(c)
(d)

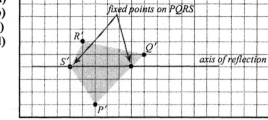

7.

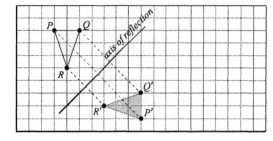

9.

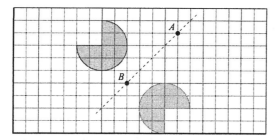

Since points *A* and *B* are fixed points, the axis of reflection must pass through these points.

B.   Rotations

11. (a) *I.* Think of *A* as the center of a clock in which *B* is the "9" and *I* is the "12."

   (b) *E.* Think of *A* as the center of a clock in which *B* is the "9" and *E* is the "3."

   (c) *G.* Think of *B* as the center of a clock in which *A* is the "3" and *G* is the "6."

   (d) *A.* Think of *B* as the center of a clock in which *D* is the "1" and *A* is the "3."

   (e) *F.* Think of *B* as the center of a clock in which *D* is the "1" and *F* is the "5."

   (f) *C.* Think of *B* as the center of a clock in which *D* is the "1" and *C* is the "9." Remember that you are rotating counterclockwise in this problem.

   (g) *E.* Think of *A* as the center of a clock in which *I* is the "12." Rotating 3690° has the same effect as rotating $10 \times 360° + 90°$. That is, as rotating 10 times around the circle and then another 90°.

   (h) *D.* Rotating 7530° has the same effect as rotating $20 \times 360° + 330°$. That is, as rotating 20 times around the circle and then another 330°.

13. (a) $360° - 250° = 110°$

   (b) $710° - 360° = 350°$

   (c) $2(360°) - 710° = 10°$

   (d) $360° \overline{)3681°}^{\ 10}$ with a remainder of 81°. Hence, a clockwise rotation of 3681° is equivalent to a clockwise rotation of 81°.

15. (a)
(b)

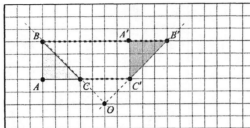

Since *BB'* and *CC'* are parallel, the intersection of *BC* and *B'C'* locates the rotocenter *O*. This is a 90° clockwise rotation.

17. (a)
(b)
(c)

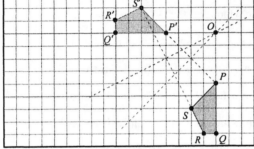

The rotocenter *O* is located at the intersection of the perpendicular bisectors to *PP'* and *SS'*. This is a 90° counterclockwise rotation.

19. (a)
(b)

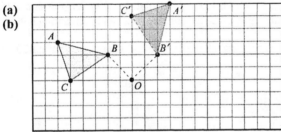

The rotocenter *O* is located somewhere on the perpendicular bisector of *BB'*. Since the 90° rotation is clockwise, *O* is located below *BB'* rather than above *BB'*.

C.   Translations

21. (a) *C*. Vector v_1 translates a point 4 units to the right so the image of *P* is *C*.

(b) *C*. Vector v_2 translates a point 4 units to the right so the image of *P* is *C*.

(c) *A*. Vector v_3 translates a point up 2 units and right 1 unit so the image of *P* is *A*.

(d) *D*. Vector v_4 translates a point down 2 units and left 1 unit so the image of *P* is *D*.

23. (a)
(b)

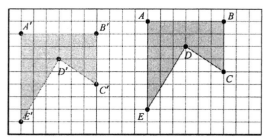

25. One possible answer: One translation is right (East) 4 miles. A second translation is up (North) 3 miles (see figure).

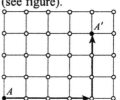

D.   Glide Reflections

27.

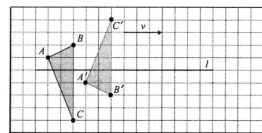

First, glide the figure 3 units to the right. Then, reflect the result about the axis *l*.

29. (a)
(b)
(c)

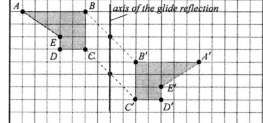

31. (a)
(b)

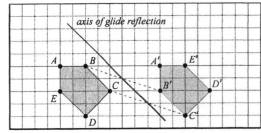

33. (a)
(b)

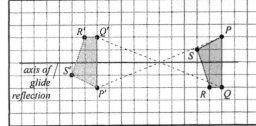

The axis of reflection cannot be determined by the midpoints of *PP'* and *QQ'* (since they are the same point). So, the axis of reflection is the line perpendicular to *PQ*.

E.   Symmetries of Finite Shapes

35. (a) Reflection with axis going through the midpoints of *AB* and *DC*; reflection with axis going through the midpoints of *AD* and *BC*; rotations of 180° and 360° with rotocenter the center of the rectangle.

  (b) No Reflections. Rotations of 180° and 360° with rotocenter the center of the parallelogram.

  (c) Reflection with axis going through the midpoints of *AB* and *DC*; rotation of 360° with rotocenter the center of the trapezoid.

37. (a) Reflections (three of them) with axis going through pairs of opposite vertices; reflections (three of them) with axis going through the midpoints of opposite sides of the hexagon; rotations of 60°, 120°, 180°, 240°, 300°, 360° with rotocenter the center of the hexagon.

  (b) Reflections with axis *AD, GJ, BE, HK, CF, IL*; rotations of 60°, 120°, 180°, 240°, 300°, 360° with rotocenter the center of the star.

39. (a) D_2; the figure has exactly 2 reflections and 2 rotations.

  (b) Z_2; the figure has no reflections and exactly 2 rotations.

  (c) D_1; the figure has exactly 1 reflection and 1 rotation.

41. (a) D_6; the figure has exactly 6 reflections and 6 rotations.

(b) D_6 ; the figure has exactly 6 reflections and 6 rotations.

43. (a) D_1 ; the letter A has exactly 1 reflection (vertical) and 1 rotation (identity).

(b) D_1 ; the letter D has exactly 1 reflection (horizontal) and 1 rotation (identity).

(c) Z_1 ; the letter L has no reflection and exactly 1 rotation (identity).

(d) Z_2 ; the letter N has no reflection and exactly 2 rotations (identity and 180°).

(e) D_1 ; the Greek letter Ω has exactly 1 reflection and 1 rotation (identity).

(f) D_2 ; the Greek letter Φ has exactly 2 reflections (vertical and horizontal) and 2 rotations (180° and the identity).

45. Answers will vary.

(a) Since symmetry type Z_1 has no reflections and exactly 1 rotation, the capital letter J is an example of this symmetry type.

(b) Since symmetry type D_1 has exactly 1 reflection and 1 rotation, the capital letter T is an example of this symmetry type.

(c) Since symmetry type Z_2 has no reflection and exactly 2 rotations, the capital letter Z is an example of this symmetry type.

(d) Since symmetry type D_2 has exactly 2 reflections and 2 rotations, the capital letter I is an example of this symmetry type.

47. Answers will vary.

(a) Symmetry type D_5 is common among many types of flowers (daisies, geraniums, etc.). The only requirements are that the flower have 5 equal, evenly spaced petals and that the petals have a reflection symmetry along their long axis. In the animal world, symmetry type D_5 is less common, but it can be found among certain types of starfish, sand dollars, and in some single celled organisms called diatoms.

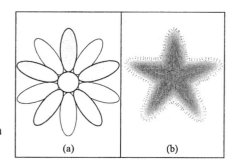

(a)   (b)

(b) The Chrysler Corporation logo is a classic example of a shape with symmetry D_5. Symmetry type D_5 is also common in automobile wheels and hubcaps. One of the largest and most unusual buildings in Washington, DC has symmetry of type D_5.

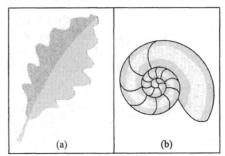

(c) Objects with symmetry type Z_1 are those whose only symmetry is the identity. Thus, any "irregular" shape fits the bill. Tree leaves, seashells, plants, and rocks more often than not have symmetry type Z_1.

(d) Examples of manmade objects with symmetry of type Z_1 abound.

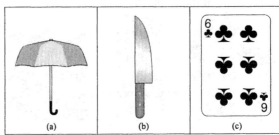

F. Symmetries of Border Patterns

49. (a) $m1$; the border pattern has translation symmetry and vertical reflection but does not have horizontal reflection, half-turn rotation, or glide reflection.

(b) $1m$; the border pattern has translation symmetry and horizontal reflection but does not have vertical reflection, half-turn, or glide reflection.

(c) 12; the border pattern has translation symmetry and half-turn rotation but does not have horizontal reflection, vertical reflection, or glide reflection, its symmetry type is.

(d) 11; the border pattern has translation symmetry but does not have horizontal reflection, vertical reflection, half-turn rotation, or glide reflection.

51. (a) $m1$; the border pattern has translation symmetry and vertical reflection but does not have horizontal reflection, glide reflection, or half-turn rotation.

(b) 12; the border pattern has translation symmetry and half-turn rotation but does not have horizontal reflection, vertical reflection, or glide reflection.

   (c) 1*g*; the border pattern has translation symmetry and glide reflection but does not have horizontal reflection, vertical reflection, or half-turn rotation.

   (d) *mg*; the border pattern has translation symmetry, vertical translation, half-turn rotation, and glide reflection, but does not have horizontal reflection.

53. 12; see, for example, 49(c).

G.   **Miscellaneous**

55. Since every proper rigid motion is equivalent to either a rotation or a translation, and a translation has no fixed points, the specified rigid motion must be equivalent to a rotation.

57. **(a)** *D*. The reflection of *P* about l_1 is the point *F*. The reflection of *F* about line l_2 is *D*. So, point *D* is the image of point *P* under the product of these reflections.

   (b) *D*. The reflection of *P* about l_2 is the point *A*. The reflection of *A* about line l_1 is *D*. So, again, point *D* is the image of point *P* under the product of these reflections.

   (c) *B*. The reflection of *P* about l_2 is the point *A*. The reflection of *A* about line l_3 is *B*.

   (d) *E*. The reflection of *P* about l_3 is the point *C*. The reflection of *C* about line l_2 is *E*. [Note that (a), (b), (c) and (d) suggest that the order in which reflections occur *sometimes* makes a difference.]

   (e) *G*. The reflection of *P* about l_1 is the point *F*. The reflection of *F* about line l_4 is *G*.

59. **(a)** When a proper rigid motion is combined with an improper rigid motion, the result is an improper rigid motion; that is, the left-right and clockwise-counterclockwise orientations on the final figure will be the reverse of the original figure.

   (b) When an improper rigid motion is combined with an improper rigid motion, the result is a proper rigid motion; that is, the left-right and clockwise-counterclockwise orientations on the final figure will be the same as the original figure. This occurs because the orientations are reversed and then reversed again. The second reversal brings the orientation back to the original orientation.

   (c) improper: this is an improper rigid motion combined with a proper rigid motion (see part a).

   (d) proper: this is an improper rigid motion combined with an improper rigid motion (see part b).

61. The combination of two improper rigid motions is a proper rigid motion. Since *C* is a fixed point, the rigid motion must be a rotation with rotocenter *C*.

JOGGING

63. **(a)** The result of applying the reflection with axis l_1, followed by the reflection with axis l_2, is a clockwise rotation with center *C* and angle of rotation $\lambda + \lambda + \beta + \beta = 2(\lambda + \beta) = 2\alpha$. One example is shown below.

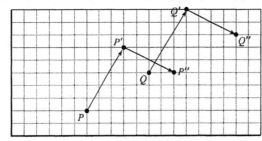

(b) The result of applying the reflection with axis l_2, followed by the reflection with axis l_1, is a counter-clockwise rotation with center C and angle of rotation 2α.

65. (a)

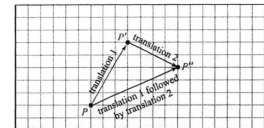

(b)

67. (a) By definition, a border has translation symmetries in exactly one direction (let's assume the horizontal direction). If the pattern had a reflection symmetry along an axis forming 45° with the horizontal direction, there would have to be a second direction of translation symmetry (vertical).

(b) If a pattern had a reflection symmetry along an axis forming an angle of $\alpha°$ with the horizontal direction, it would have to have translation symmetry in a direction that forms an angle of $2\alpha°$ with the horizontal. This could only happen for $\alpha = 90°$ or $\alpha = 180°$ (since the only allowable direction for translation symmetries is the horizontal).

69. (a)

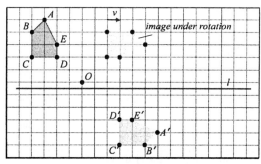

(b) Rotations and translations are proper rigid motions, and hence preserve clockwise-counterclockwise orientations. The given motion is an improper rigid motion (it reverses the clockwise-counterclockwise orientation). Since performing the same rotation and glide reflection again results in the original figure, this is a reflection (rather than a glide reflection) whose axis is shown in the figure below.

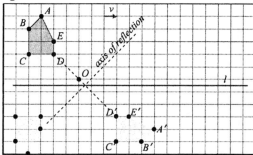

[The figure in the bottom left is the result of rotation $A'B'C'D'E'$ 90° about O. Performing the glide reflection on that figure results in $ABCDE$.]

71. Rotations and translations are proper rigid motions, and hence preserve clockwise-counterclockwise orientations. The given motion is an improper rigid motion (it reverses the clockwise-counterclockwise orientation). If the rigid motion was a reflection, then PP', RR', and QQ' would all be perpendicular to the axis of reflection and hence would all be parallel. It must be a glide reflection (the only rigid motion left).

Chapter 12

WALKING

A.   The Koch Snowflake and Variations

1.

	Start	Step 1	Step 2	Step 3	Step 4	...	Step 30	Step N
Number of sides	3	12	48	192	768		3×4^{30}	3×4^{N}
Length of each side	1 in	1/3 in	1/9 in	1/27 in	1/81 in		$\dfrac{1}{3^{30}}$ in	$\dfrac{1}{3^{N}}$ in
Length of the boundary	3 in	4 in	48/9 in	192/27 in	768/81 in		≈ 0.265 mi	$3 \times \left(\dfrac{4}{3}\right)^{N}$ in

The number of sides increases by a factor of 4 at each step since each side is replaced by 4 sides of 1/3 the length. The length of the boundary can be found by multiplying the number of sides at step N by the length of each side at step N.

3.

	Start	Step 1	Step 2	Step 3	Step 4
Area	24 in^2	32 in^2	$\dfrac{320}{9}$ in^2	$\dfrac{3,008}{81}$ in^2	$\dfrac{27,584}{729}$ in^2

At step 1, three triangles are added each having 1/9 the area of the original triangle. That is, the area at step 1 is $24 \text{ in}^2 + \dfrac{3 \times 24}{9} \text{ in}^2 = 32 \text{ in}^2$.

At step 2, $3 \times 4 = 12$ triangles are added each having $\dfrac{1}{9^2}$ the area of the original triangle. That is, the area at step 2 is $24 \text{ in}^2 + \dfrac{3 \times 24}{9} \text{ in}^2 + \dfrac{12 \times 24}{9^2} \text{ in}^2 = \dfrac{2,880}{81} \text{ in}^2 = \dfrac{320}{9} \text{ in}^2$.

At step 3, $3 \times 4^2 = 48$ triangles are added each having $\dfrac{1}{9^3}$ the area of the original triangle. That is, the area at step 3 is $24 \text{ in}^2 + \dfrac{3 \times 24}{9} \text{ in}^2 + \dfrac{3 \times 4 \times 24}{9^2} \text{ in}^2 + \dfrac{3 \times 4^2 \times 24}{9^3} \text{ in}^2 = \dfrac{27,072}{729} \text{ in}^2 = \dfrac{3,008}{81} \text{ in}^2$.

At step 4, $3 \times 4^3 = 192$ triangles are added each having $\dfrac{1}{9^4}$ the area of the original triangle. That is, the area at step 4 is $24 \text{ in}^2 + \dfrac{3 \times 24}{9} \text{ in}^2 + \dfrac{3 \times 4 \times 24}{9^2} \text{ in}^2 + \dfrac{3 \times 4^2 \times 24}{9^3} \text{ in}^2 + \dfrac{3 \times 4^3 \times 24}{9^4} \text{ in}^2 = \dfrac{27,584}{729} \text{ in}^2$.

5. $1.6 \times 24 = 38.4 \text{ in}^2$

7.

	Start	Step 1	Step 2	Step 3	Step 4	...	Step 50	Step N
Number of sides	4	20	100	500	2500		4×5^{50}	4×5^{N}
Length of each side	1 in	1/3 in	1/9 in	1/27 in	1/81 in		$\dfrac{1}{3^{50}}$ in	$\dfrac{1}{3^{N}}$ in
Length of the boundary	4 in	20/3 in	100/9 in	500/27 in	2500/81 in		$\approx 7{,}810{,}562$ mi	$4 \times \left(\dfrac{5}{3}\right)^{N}$

The number of sides increases by a factor of 5 at each step since each side is replaced by 5 sides of 1/3 the length. The length of the boundary can be found by multiplying the number of sides at step N by the length of each side at step N.

9.

	Start	Step 1	Step 2	Step 3	Step 4
Area	1 in^2	$\dfrac{13}{9}$ in^2	$\dfrac{137}{81}$ in^2	$\dfrac{1333}{729}$ in^2	$\dfrac{12{,}497}{6561}$ in^2

At step 1, four squares are added each having 1/9 the area of the original square. That is, the area at step 1 is $1 \text{ in}^2 + \dfrac{4 \times 1}{9} \text{ in}^2 = \dfrac{13}{9} \text{ in}^2$.

At step 2, $4 \times 5 = 20$ squares are added each having $\dfrac{1}{9^2}$ the area of the original square. That is, the area at step 2 is $1 \text{ in}^2 + \dfrac{4 \times 1}{9} \text{ in}^2 + \dfrac{20 \times 1}{9^2} \text{ in}^2 = \dfrac{137}{81} \text{ in}^2$.

At step 3, $4 \times 5^2 = 100$ squares are added each having $\dfrac{1}{9^3}$ the area of the original square. That is, the area at step 3 is $1 \text{ in}^2 + \dfrac{4 \times 1}{9} \text{ in}^2 + \dfrac{4 \times 5 \times 1}{9^2} \text{ in}^2 + \dfrac{4 \times 5^2 \times 1}{9^3} \text{ in}^2 = \dfrac{1333}{729} \text{ in}^2$.

At step 4, $4 \times 5^3 = 500$ squares are added each having $\dfrac{1}{9^4}$ the area of the original square. That is, the area at step 4 is $1 \text{ in}^2 + \dfrac{4 \times 1}{9} \text{ in}^2 + \dfrac{4 \times 5 \times 1}{9^2} \text{ in}^2 + \dfrac{4 \times 5^2 \times 1}{9^3} \text{ in}^2 + \dfrac{4 \times 5^3 \times 1}{9^4} \text{ in}^2 = \dfrac{12{,}497}{6561} \text{ in}^2$.

11.

	Start	Step 1	Step 2	Step 3	Step 4	...	Step 40	Step N
Number of sides	3	12	48	192	768		3×4^{40}	3×4^N
Length of each side	a	$a/3$	$a/9$	$a/27$	$a/81$		$\dfrac{a}{3^{40}}$	$\dfrac{a}{3^N}$
Length of the boundary	$3a$	$4a$	$\dfrac{48}{9}a$	$\dfrac{192}{27}a$	$\dfrac{768}{81}a$		$\dfrac{3 \times 4^{40}}{3^{40}}a$	$3a \times \left(\dfrac{4}{3}\right)^N$

The number of sides increases by a factor of 4 at each step since each side is replaced by 4 sides of 1/3 the length. The length of the boundary can be found by multiplying the number of sides at step N by the length of each side at step N.

13.

	Start	Step 1	Step 2	Step 3	Step 4	Step 5
Area	81 in^2	54 in^2	42 in^2	$\dfrac{110}{3} \text{ in}^2$	$\dfrac{926}{27} \text{ in}^2$	$\dfrac{8,078}{243} \text{ in}^2$

At step 1, three triangles are subtracted each having 1/9 the area of the original triangle. That is, the area at step 1 is $81 \text{ in}^2 - \dfrac{3 \times 81}{9} \text{ in}^2 = 54 \text{ in}^2$.

At step 2, $3 \times 4 = 12$ triangles are subtracted each having $\dfrac{1}{9^2}$ the area of the original triangle. That is, the

area at step 2 is $81 \text{ in}^2 - \dfrac{3 \times 81}{9} \text{ in}^2 - \dfrac{12 \times 81}{9^2} \text{ in}^2 = 42 \text{ in}^2$.

At step 3, $3 \times 4^2 = 48$ triangles are subtracted each having $\dfrac{1}{9^3}$ the area of the original triangle. That is,

the area at step 3 is $81 \text{ in}^2 - \dfrac{3 \times 81}{9} \text{ in}^2 - \dfrac{3 \times 4 \times 81}{9^2} \text{ in}^2 - \dfrac{3 \times 4^2 \times 81}{9^3} \text{ in}^2 = \dfrac{26,730}{729} \text{ in}^2 = \dfrac{110}{3} \text{ in}^2$.

At step 4, $3 \times 4^3 = 192$ triangles are subtracted each having $\dfrac{1}{9^4}$ the area of the original triangle. That is,

the area at step 4 is $81 \text{ in}^2 - \dfrac{3 \times 81}{9} \text{ in}^2 - \dfrac{3 \times 4 \times 81}{9^2} \text{ in}^2 - \dfrac{3 \times 4^2 \times 81}{9^3} \text{ in}^2 - \dfrac{3 \times 4^3 \times 81}{9^4} \text{ in}^2$

$= \dfrac{225,018}{6,561} \text{ in}^2 = \dfrac{2,778}{81} \text{ in}^2 = \dfrac{926}{27} \text{ in}^2$.

At step 5, $3 \times 4^4 = 768$ triangles are subtracted each having $\dfrac{1}{9^5}$ the area of the original triangle. That is, the area at step 5 is $81 \text{ in}^2 - \dfrac{3 \times 81}{9} \text{ in}^2 - \dfrac{3 \times 4 \times 81}{9^2} \text{ in}^2 - \dfrac{3 \times 4^2 \times 81}{9^3} \text{ in}^2 - \dfrac{3 \times 4^3 \times 81}{9^4} \text{ in}^2 - \dfrac{3 \times 4^4 \times 81}{9^5} \text{ in}^2$

$= \dfrac{1{,}962{,}954}{59{,}049} \text{ in}^2 = \dfrac{24{,}234}{729} \text{ in}^2 = \dfrac{8{,}078}{243} \text{ in}^2 .$

15.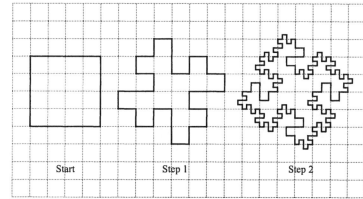

17. (a)

	Start	Step 1	Step 2	Step 3	Step 4
Area	A	A	A	A	A

At each step, the area added is the same as the area subtracted.

(b) If the area of the starting square is A, then the area of the resulting quadratic Koch island is also A (what the rule giveth, the rule taketh away).

B.   The Sierpinski Gasket and Variations

19.

	Start	Step 1	Step 2	Step 3	Step 4	...	Step N
Area	1	$\dfrac{3}{4}$	$\left(\dfrac{3}{4}\right)^2$	$\left(\dfrac{3}{4}\right)^3$	$\left(\dfrac{3}{4}\right)^4$		$\left(\dfrac{3}{4}\right)^N$

At each step of construction, ¾ of each triangle remains.

21. The area of the Sierpinski gasket is smaller than the area of the gasket formed during any step of construction. That is, if the area of the original triangle is 1, then the area of the Sierpinski gasket is less than $\left(\dfrac{3}{4}\right)^N$ for every positive value of N. Since $0 < 3/4 < 1$, the value of $\left(\dfrac{3}{4}\right)^N$ can be made smaller than any positive quantity for a large enough choice of N. It follows that the Sierpinski gasket can also be made smaller than any positive quantity.

23.

	Start	Step 1	Step 2	Step 3	Step 4	...	Step N
Number of triangles	1	3	9	27	81		3^N
Perimeter of each triangle	24 in	12 in	6 in	3 in	1.5 in		$24 \times \left(\dfrac{1}{2}\right)^N$ in
Length of the boundary	24 in	36 in	54 in	81 in	121.5 in		$24 \times \left(\dfrac{3}{2}\right)^N$ in

25.

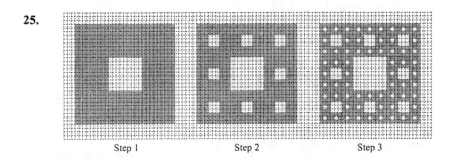

Step 1   Step 2   Step 3

27. (a)

	Start	Step 1	Step 2	Step 3	Step 4
Perimeter	4	16/3	80/9	496/27	3536/81

Note that the boundary will include the outside perimeter and the perimeters of all removed squares.

We also note that each removed square will have sides of length $\dfrac{1}{3}$ that of the square it was

removed from. Since each side of the original square has length 1, we have:

Start: $4 \cdot 1 = 4$;

Step 1: $4 + 4 \cdot \dfrac{1}{3} = 4 + \dfrac{4}{3} = \dfrac{12}{3} + \dfrac{4}{3} = \dfrac{16}{3}$

(Previous boundary plus boundary of new hole with sides of length $\dfrac{1}{3}$)

Step 2: $\dfrac{16}{3} + 8 \cdot \dfrac{4}{9} = \dfrac{80}{9}$

(Previous boundary plus boundary of 8 new holes, each with sides of length $\dfrac{1}{3} \cdot \dfrac{1}{3} = \dfrac{1}{9}$)

Step 3: $\dfrac{80}{9} + 8^2 \cdot \dfrac{4}{27} = \dfrac{496}{27}$

(Previous boundary plus boundary of 8^2 new holes, each with sides of length $\frac{1}{3} \cdot \frac{1}{3} \cdot \frac{1}{3} = \frac{1}{27}$)

Step 4: $\frac{496}{27} + 8^3 \cdot \frac{4}{81} = \frac{3536}{81}$

(Previous boundary plus boundary of 8^3 new holes, each with sides of length $\left(\frac{1}{3}\right)^4 = \frac{1}{81}$)

(b) At step $N+1$, there are 8^N new holes, each with sides of length $\left(\frac{1}{3}\right)^{N+1} = \frac{1}{3^{N+1}}$ so, perimeter of the

"carpet" at step $N + 1$ is $L + 8^N \cdot \frac{4}{3^{N+1}}$.

29.

	Start	Step 1	Step 2	Step 3	Step 4	...	Step N
Number of triangles	1	6	$6^2 = 36$	$6^3 = 216$	$6^4 = 1296$		6^N
Area of each triangle	A	$\frac{1}{9}A$	$\frac{1}{9^2}A$	$\frac{1}{9^3}A$	$\frac{1}{9^4}A$		$\frac{1}{9^N}A$
Area of gasket	A	$\frac{2}{3}A$	$\left(\frac{2}{3}\right)^2 A$	$\left(\frac{2}{3}\right)^3 A$	$\left(\frac{2}{3}\right)^4 A$		$\left(\frac{2}{3}\right)^N A$

The area of the gasket at step N of the construction will be the product of the number of triangles at that step and the area of each such triangle. The length of each side of each triangle in step N is 1/3 of the length of a side in step N-1. So, the area at step N is 1/9 that of the area of a triangle in step N-1.

31. The area of the Sierpinski ternary gasket is smaller than the area of the gasket formed during any step of construction. That is, if the area of the original triangle is A, then the area of the Sierpinski ternary gasket is less than $\left(\frac{2}{3}\right)^N A$ for every positive value of N. Since $0 < 2/3 < 1$, the value of $\left(\frac{2}{3}\right)^N A$ can be made smaller than any positive quantity for a large enough choice of N. It follows that the Sierpinski ternary gasket can also be made smaller than any positive quantity.

C.   The Chaos game and Variations

33. The coordinates of each point are:
P_1 : (32, 0);
P_2 : (16, 0), the midpoint of A and P_1;
P_3 : (8, 16), the midpoint of C and P_2;
P_4 : (20, 8), the midpoint of B and P_3;
P_5 :(10, 20), the midpoint of C and P_4;

P_6 : (5, 26), the midpoint of C and P_5.

$C(0,32)$

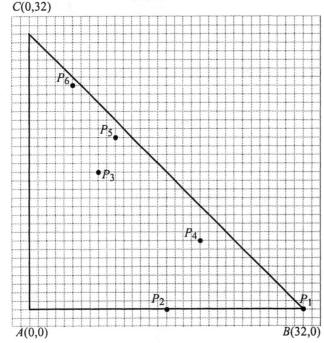

35. P_3 : (8, 0), the midpoint of A and P_2;

P_4 : (20, 0), the midpoint of B and P_3;

P_5 : (10, 16), the midpoint of C and P_4;

P_6 : (5, 24), the midpoint of C and P_5.

37. General note: Find new coordinate by picking new x-value to be $\dfrac{2}{3}$ of way from x-coordinate of first point to x-coordinate of second point and picking new y-value to be $\dfrac{2}{3}$ of way from y-coordinate of first point to y-coordinate of second point.

(a) The coordinates of each point are:

P_1 : $(0, 27)$;

P_2 : $(18, 9)$, $\dfrac{2}{3}$ of way from P_1 to B;

P_3 : $(6, 3)$, $\dfrac{2}{3}$ of way from P_2 to A;

P_4 : $(20, 1)$, $\dfrac{2}{3}$ of way from P_3 to B.

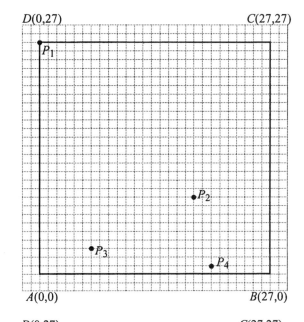

(b) The coordinates of each point are:

P_1 : $(27, 27)$;

P_2 : $(27, 9)$, $\dfrac{2}{3}$ of way from P_1 to B;

P_3 : $(9, 3)$, $\dfrac{2}{3}$ of way from P_2 to A;

P_4 : $(21, 1)$, $\dfrac{2}{3}$ of way from P_3 to B.

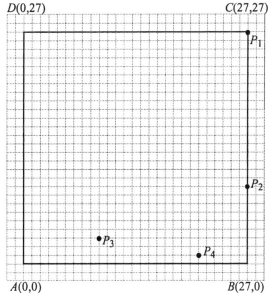

(c) The coordinates of each point are:

P_1 : (27, 27);

P_2 : (27, 27);

P_3 : (9, 9), $\frac{2}{3}$ of way from P_2 to A;

P_4 : (3, 3), $\frac{2}{3}$ of way from P_3 to A.

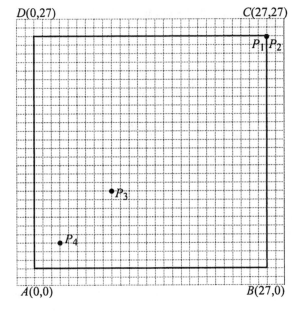

39. (a) P_1 : (0, 27); P_2 : (18, 9), $\frac{2}{3}$ of way from P_1 to B;

P_3 : (6, 3), $\frac{2}{3}$ of way from P_2 to A; P_4 : (20, 1), $\frac{2}{3}$ of way from P_3 to B.

(b) P_1 : (27, 27); P_2 : (9, 9), $\frac{2}{3}$ of way from P_1 to A;

P_3 : (3, 3), $\frac{2}{3}$ of way from P_2 to A; P_4 : (19, 19), $\frac{2}{3}$ of way from P_3 to C.

(c) P_1 : (0, 0); P_2 : (18, 18), $\frac{2}{3}$ of way from P_1 to C;

P_3 : (6, 24), $\frac{2}{3}$ of way from P_2 to D; P_4 : (20, 8), $\frac{2}{3}$ of way from P_3 to B.

D.   Operations with Complex Numbers

41.   (a) $(1+i)^2 + (1+i) = 1 + 2i - 1 + (1+i)$
$$= 1 + 3i$$

(b) $(1-i)^2 + (1-i) = 1 - 2i - 1 + (1-i)$
$$= 1 - 3i$$

(c) $(-1+i)^2 + (-1+i) = 1 - 2i - 1 + (-1+i)$
$$= -1 - i$$

43.   (a) $(-0.25 + 0.25i)^2 + (-0.25 + 0.25i)$
$$= 0.0625 - 0.125i - 0.0625 + (-0.25 + 0.25i)$$
$$= -0.25 + 0.125i$$

(b) $(-0.25 - 0.25i)^2 + (-0.25 - 0.25i)$
$$= 0.0625 + 0.125i - 0.0625 + (-0.25 - 0.25i)$$
$$= -0.25 - 0.125i$$

45.   (a) Since the value of $i(1+i) = -1 + i$,
$i^2(1+i) = -1 - i$, and $i^3(1+i) = 1 - i$ we
plot $1+i$, $-1+i$, $-1-i$, and $1-i$. This
is just like plotting (1,1), (-1,1), (-1,-1),
and (1,-1).

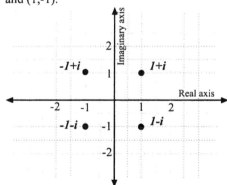

(b) Since the value of $i(3-2i) = 2 + 3i$,
$i^2(3-2i) = -3 + 2i$, and
$i^3(3-2i) = -2 - 3i$ we plot $3 - 2i$,

$2 + 3i$, $-3 + 2i$, and $-2 - 3i$.

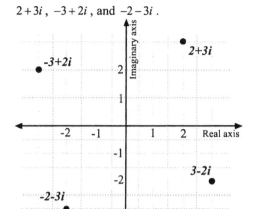

(c) The effect of multiplying each point in
(a) and (b) by i is a 90-degree
counterclockwise rotation.

E.   Mandelbrot Sequences

47.   (a) $s_1 = (-2)^2 + (-2) = 4 - 2 = 2;$
$s_2 = (2)^2 + (-2) = 4 - 2 = 2;$
$s_3 = (2)^2 + (-2) = 4 - 2 = 2;$
$s_4 = (2)^2 + (-2) = 4 - 2 = 2.$

(b) The sequence is attracted to 2, so
$s_{100} = 2$.

(c) Since each number in the sequence is 2,
the sequence is attracted to 2.

49.   (a) $s_1 = (-0.5)^2 + (-0.5) = -0.25;$
$s_2 = (-0.25)^2 + (-0.5) = -0.4375;$
$s_3 = (-0.4375)^2 + (-0.5) \approx -0.3086;$
$s_4 = (-0.3086)^2 + (-0.5) \approx -0.4048;$
$s_5 = (-0.4048)^2 + (-0.5) \approx -0.3361.$

(b) $s_{N+1} = (-0.366)^2 + (-0.5) \approx -0.366$

(c) From part (b), we see that $s_N = s_{N+1}$, so
the sequence must be attracted to $-$
0.3660 (rounded to 4 decimal places).

51. (a) $s_1 = \left(\dfrac{1}{2}\right)^2 + \dfrac{1}{2} = \dfrac{1}{4} + \dfrac{1}{2} = \dfrac{3}{4}$;

$s_2 = \left(\dfrac{3}{4}\right)^2 + \dfrac{1}{2} = \dfrac{9}{16} + \dfrac{1}{2} = \dfrac{17}{16}$;

$s_3 = \left(\dfrac{17}{16}\right)^2 + \dfrac{1}{2} = \dfrac{289}{256} + \dfrac{1}{2} = \dfrac{417}{256}$.

(b) If $s_N > 1$, we have $(s_N)^2 > s_N$, and so

$(s_N)^2 + \dfrac{1}{2} > s_N + \dfrac{1}{2}$. But

$(s_N)^2 + \dfrac{1}{2} = s_{N+1}$, so we have

$s_{N+1} > s_N + \dfrac{1}{2} > s_N$.

(c) Notice from part (a) that $s_2 > 1$. From part (b) this implies that the sequence is escaping.

53. (a) $s_{N+1} = (s_N)^2 + s$

$38 = (6)^2 + s$

$38 = 36 + s$

$s = 2$

(b) $s_1 = (2)^2 + 2$

$= 4 + 2$

$= 6$;

$N = 1$

JOGGING

55. At each step, one new hole is introduced for each solid triangle and the number of solid triangles is tripled (multiplied by three).

Step	Holes	Solid Triangles
1	1	3
2	$1 + 3$	3^2
3	$1 + 3 + 3^2$	3^3
4	$1 + 3 + 3^2 + 3^3$	3^4

Using the formula for the sum of the terms in a geometric sequence in which $r = 3$ (found in chapter 10), we see that at the Nth step there will be

$$1 + 3 + 3^2 + 3^3 + \cdots + 3^{N-1} = \dfrac{1 - 3^N}{1 - 3} = \dfrac{3^N - 1}{2}$$

holes.

57. At the first step, a cube is removed from each of the six faces and from the center for a total of 7 cubes removed. At the next step, each of the 20 remaining cubes has 7 cubes removed (i.e, 20×7 cubes are removed). In the second step, there are 20^2 remaining cubes each of which has 7 cubes removed (i.e. $20^2 \times 7$ cubes are removed). Continuing in this fashion, we see that $20^{N-1} \times 7$ are removed from the 20^{N-1} remaining cubes in step N.

	Number of cubes removed
Start	0
Step 1	7
Step 2	20×7
Step 3	$20^2 \times 7$
...	
Step N	$20^{N-1} \times 7$

59. (a) Reflection with axis a vertical line passing through the center of the snowflake; reflections with axes lines making 30°, 60°, 90°, 120°, 150° angles with the vertical axis of the snowflake and passing through the center of the snowflake.

(b) Rotations of 60°, 120°, 180°, 240°, 300°, 360° with rotocenter the center of the snowflake.

(c) D_6

61. (a) Using the formula for the sum of the terms in a geometric sequence in which $r = 4/9$ (found in chapter 10), we have

$$1 + \left(\tfrac{4}{9}\right) + \left(\tfrac{4}{9}\right)^2 + \cdots + \left(\tfrac{4}{9}\right)^{N-1}$$

$$= \frac{1 - \left(\tfrac{4}{9}\right)^N}{\left(\tfrac{5}{9}\right)} = \tfrac{9}{5}\left[1 - \left(\tfrac{4}{9}\right)^N\right]$$

(b) Using the result in (a) we have

$$\left(\frac{A}{3}\right) + \left(\frac{A}{3}\right)\left(\frac{4}{9}\right) + \left(\frac{A}{3}\right)\left(\frac{4}{9}\right)^2 + \ldots + \left(\frac{A}{3}\right)\left(\frac{4}{9}\right)^{N-1} =$$

$$\left(\frac{A}{3}\right)\left[1 + \left(\frac{4}{9}\right) + \left(\frac{4}{9}\right)^2 + \ldots + \left(\frac{4}{9}\right)^{N-1}\right] =$$

$$\left(\frac{A}{3}\right)\left(\frac{9}{5}\right)\left[1 - \left(\frac{4}{9}\right)^N\right] = \left(\frac{3}{5}\right)A\left[1 - \left(\frac{4}{9}\right)^N\right].$$

63. If it is attracted to a number, then we have $s_{N+1} = (s_N)^2 + (0.2)$ and we set $s_{N+1} = s_N$. Substituting, we have $s_N = (s_N)^2 + 0.2$ so $s_N{}^2 - s_N + 0.2 = 0$. Solve via the Quadratic Equation: $s_N = \dfrac{1 \pm \sqrt{1 - 4(0.2)}}{2} = \dfrac{1 \pm \sqrt{0.2}}{2}.$

Solving yields $s_N = \dfrac{1 + \sqrt{0.2}}{2}$ or $s_N = \dfrac{1 - \sqrt{0.2}}{2}.$ We proceed to examine a few terms of the sequence:
Step 10: 0.276241069;
Step 20: 0.2763927972;
Step 30: 0.2763932012.
This sequence is attracted to
$\dfrac{1 - \sqrt{0.2}}{2} \approx 0.276393....$

65. The first twenty steps: 0.3125, -1.15234375, 0.0778961182, -1.24393219, 0.297367305,-

1.16157269, 0.0992511044, -1.24014922, 0.287970084, -1.16707323, 0.112059926, -1.23744257, 0.281264121, -1.17089049, 0.120984549, -1.23536274, 0.276121097, -1.17375714, 0.127705824, -1.23369122.
The numbers oscillate, alternating between positive and negative values, but always staying within the interval $-1.25 \le x \le 0.3125$. The sequence is periodic.

Chapter 13

WALKING

A.   Surveys and Polls

1. **(a)** The population for this survey is the gumballs in the jar.

    (b) The sample for this survey is the 25 gumballs draw out of the jar.

    (c) 32%
    The proportion of red gumballs in the sample is $\frac{8}{25} = 0.32$.

    (d) The sampling method used for this survey was simple random sampling. Each set of n gumballs had the same probability of being selected as any other set of n gumballs.

3. **(a)** 25%
    The parameter for the proportion of red gumballs in the jar is $\frac{50}{200} = 0.25$.

    (b) 7%
    The sampling error is 32% - 25% = 7%.

    (c) Sampling variability.
    Simple random sampling was used. This survey method does not suffer from selection bias.

5. **(a)** The sampling proportion for this survey is $\frac{680}{8325} \approx 0.082$; that is, approximately 8.2%.

    (b) 306/680 = 45%

7. Of the people surveyed, 45% indicated they would vote for Smith $\left(\frac{306}{680} = 0.45\right)$. The actual percentage was 42%. Since 45% − 42% = 3%, the sampling error for

Smith was 3%.
Similarly, the sampling error for Jones was 43% - 40% = 3% and the sampling error for Brown was 15% - 15% = 0%.

9. **(a)** The sample for this survey is the 350 students attending the Eureka High football game before the election.

    (b) $\frac{350}{1250} = 28\%$

11. **(a)** The population consists of all 1250 students at Eureka High School while the sampling frame consists only of those students that attended the football game a week prior to the election.

    (b) The sampling error is mainly a result of sampling bias. The sampling frame (and hence any sample taken from it) is not representative of the population. Students that choose to attend a football game are not representative of all Eureka High students.

13. **(a)** The sampling frame is all married people who read Dear Abby's column.

    (b) Abby's target population appears to be all married people. However, she is sampling from a (non-representative) subset of the population – a sampling frame that consists of those married couples that read her column. The sampling frame is quite different than the target population.

    (c) The sample chose itself. That is, the sample was chosen via self-selection (which is a type of convenience sample).

    (d) 85% is a statistic, since it is based on data taken from a sample.

15. **(a)** $\dfrac{44,807}{60,550} = 74.0\%$

(b) $\dfrac{127,318 + 44,807}{210,336} \approx 81.8\%$

(c) These estimates are probably not very accurate. The sample was far from being representative of the entire target population. One reason is that the sampling frame is so different from the target population. Another reason is that even if the sampling frame was similar to the target population, the survey is subject to nonresponse bias.

17. **(a)** The target population of this survey is the citizens of Cleansburg.

(b) The sampling frame is limited to that part of the target population that passes by a city street corner between 4:00 p.m. and 6:00 p.m.. It excludes citizens of Cleansburg having other responsibilities during that time of day.

19. **(a)** The choice of street corner could make a great deal of difference in the responses collected.

(b) D (We are making the assumption that people who live or work downtown are much more likely to answer yes than people in other parts of town.)

(c) Yes, the survey was subject to selection bias for two main reasons. (i) People out on the street between 4 p.m. and 6 p.m. are not representative of the population at large. For example, office and white-collar workers are much more likely to be in the sample than homemakers and school teachers. (ii) The five street corners were chosen by the interviewers and the passersby are unlikely to represent a cross section of the city.

(d) No, no attempt was made to use quotas to get a representative cross section of the population.

21. **(a)** Assuming that the registrar has a complete list of the 15,000 undergraduates at Tasmania State University, the target population and the sampling frame both consist of all undergraduates at TSU.

(b) $N = 15,000$

23. **(a)** In simple random sampling, any two members of the population have as much chance of both being in the sample as any other two. But in this sample, two people with the same last name—say Len Euler and Linda Euler—have no chance of being in the sample together.

(b) Sampling variability. The students sampled appear to be a reasonably representative cross section of all TSU undergraduates that would attempt to enroll in Math 101.

25. **(a)** Stratified sampling. The trees are broken into three different strata (by variety) and then a random sample is taken from each stratum.

(b) Quota sampling. The grower is using a systematic method to force the sample to fit a particular profile. However, because the grower is human, sampling bias could then be introduced. When selecting 300 trees of variety A, the grower does not select them at random. Selecting 300 trees in one particular part of the orchard could bias the yield.

27. **(a)** Convenience sampling.
George is selecting those units in the population that are easily accessible.

(b) Census.
The evaluation is a survey of all the students in the population.

(c) Stratified sampling.
The student newspaper is dividing the population into strata and then selecting a proportionately sized random sample from each stratum.

(d) Simple random sampling.
Every subset of 3 players has the same chance of being selected as any other subset of 3 players.

(e) Quota sampling.
The coach is attempting to force the sample to fit a particular profile.

B.   The Capture-Recapture Method

29. $N = \dfrac{n_2}{k} \cdot n_1 = \dfrac{120}{30} \cdot 500 = 2000$

31. $N = \dfrac{n_2}{k} \cdot n_1 = \dfrac{28}{4} \cdot 12 = 84$ quarters.

(Note: To estimate the number of quarters, we disregard the nickels and dimes—they are irrelevant.)

33. $N = \dfrac{n_2}{k} \cdot n_1 = \dfrac{43}{8} \cdot 23 = 123.625$; Rounding

to the nearest integer gives us 124 dimes.
(Note: To estimate the number of dimes, we disregard the quarters and nickels—they are irrelevant.)

35. $N = \dfrac{n_2}{k} \cdot n_1 = \dfrac{660}{7} \cdot 1700 \approx 160,285$

Rounding to the nearest thousand gives an estimate of 160,000 lake sturgeon in the Lake of the Woods.

37. (a) $N = \dfrac{n_2}{k} \cdot n_1 = \dfrac{1540}{171} \cdot 4064 \approx 36,600$

(b) $\dfrac{1}{3}(0.89 \times 36,600) \approx 10,860$

C.   Clinical Studies

39. (a) The target population for this study is anyone who could have a cold and would consider buying vitamin X (i.e., pretty much all adults).

(b) The sampling frame is only a small portion of the target population. It only consists of college students in the San Diego area that are suffering from colds.

(c) Yes. This sample would likely under represent older adults and those living in colder climates.

41. Four different problems with this study that indicate poor design include
   (i) using college students (College students are not a representative cross section of the population in terms of age and therefore in terms of how they would respond to the treatment.),
   (ii) using subjects only from the San Diego area,
   (iii) offering money as an incentive to participate, and
   (iv) allowing self-reporting (the subjects themselves determine when their colds are over).

43. The target population for this study is all potential knee surgery patients.

45. (a) Yes, this was a controlled placebo experiment. There was one control group receiving the sham-surgery (placebo) and two treatment groups.

(b) The first treatment group consisted of those patients receiving arthroscopic debridement. The second treatment group consisted of those patients receiving arthroscopic lavage.

(c) Yes, this study could be considered a randomized controlled experiment since the 180 patients in the study were assigned to a treatment group at random.

(d) This was a blind experiment. The doctors certainly knew which surgery they were performing on each patient.

47. The professor was conducting a clinical study because he was, after all, trying to establish the connection between a cause (10 milligrams of caffeine a day) and an effect (improved performance in college courses). Other than that, the experiment had little going for it: it was not controlled (no control group); not randomized (the subjects were chosen because of their poor grades); no placebo was used and consequently the study was not double blind.

49. (a) It is likely that the study was blind but not double-blind since the professor knew who was in the study.

(b) Three possible causes that could have confounded the results of this study include the following.
(i) A regular visit to the professor's office could in itself be a boost to a student's self-confidence and help improve his or her grades.
(ii) The "individualized tutoring" that took place during the office meetings could also be the reason for improved performance.
(iii) The students selected for the study all got F's on their first midterm, making them likely candidates to show some improvement.

51. The target population consists of all people having a history of colorectal adenomas. The point of the study is to determine the effect that Vioxx had on this population. That is, whether recurrence of colorectal polyps could be prevented in this population.

53. (a) The treatment group consisted of the 1287 patients that were given 25 daily milligrams of Vioxx. The control group

consisted of the 1299 patients that were given a placebo.

(b) This is an experiment since members of the population received a treatment. It is a controlled placebo experiment since there was a control group that did not receive the treatment, but instead received a placebo. It is a randomized controlled experiment since the 2586 participants were randomly divided into the treatment and control groups. The study was double blind since neither the participants nor the doctors involved in the clinical trial knew who was in each group.

D. **Miscellaneous**

55. (a) Spurlock's study was a clinical trial since a treatment was imposed (eating three meals at McDonald's every day for 30 days) on a sample of the population.

(b) The target population is the set of "average Americans."

(c) The sample consisted of one person (that being Mr. Spurlock himself).

(d) Three problems with this study that indicate poor design include the following (there are countless others).
(i) The use of a sample that is not representative of the population.
(ii) A small sample size (1 person).
(iii) The lack of a control group in which a sample of "average Americans" curtailed their physical activity and ate the same number of calories as the treatment group.

57. (a) This study was a clinical trial since a treatment was imposed (the heavy reliance on supplemental materials, online practice exercises and interactive tutorials) on a sample of the population.

(b) Some possible confounding variables in this study include the following.
(i) The instructors used in the treatment group may be more excited about this new curricular approach or they may be better teachers.
(ii) If students in this particular intermediate algebra class were self-selected, they may not be representative of the target population.
(iii) Students in the treatment group may have benefited simply by being forced to put more time studying into the course. To eliminate this possible confounding variable, the control groups should be asked to spend the same amount of time studying out of class.

59. (a) parameter
This statement refers to the entire population of students taking the SAT math test.

(b) statistic
A sample of the population of new automobiles are crash tested.

(c) statistic
A sample of the population of Mr. Johnson's blood tested positive.

(d) statistic
The poll did not sample all Americans; the statement refers to a sample of the population.

JOGGING

61. (a) The populations are (i) the entire sky; (ii) all the coffee in the cup; (iii) the entire Math 101 class.

(b) In none of the three examples is the sample random.

(c) (i) In some situations one can have a good idea as to whether it will rain or not by seeing only a small section of the sky, but in many other situations rain

clouds can be patchy and one might draw the wrong conclusions by just peeking out the window. (ii) If the coffee is burning hot on top, it is likely to be pretty hot throughout, so Betty's conclusion is likely to be valid. (iii) Since Carla used convenience sampling and those students sitting next to her are not likely to be a representative sample, her conclusion is likely to be invalid.

63. (a) The results of this survey might be invalid because the question was worded in a way that made it almost impossible to answer yes.

(b) "Will you support some form of tax increase if it can be proven to you that such a tax increase is justified?" is better, but still not neutral. "Do you support or oppose some form of tax increase?" is bland but probably as neutral as one can get.

(c) Many such examples exist. A very real example is a question such as "Would you describe yourself as pro-life or not?" or "Would you describe yourself as pro-choice or not?"

65. (a) Under method 1, people whose phone numbers are unlisted are automatically ruled out from the sample. At the same time, method 1 is cheaper and easier to implement than method 2. Method 2 will typically produce more reliable data since the sample selected would better represent the population.

(b) For this particular situation, method 2 is likely to produce much more reliable data than method 1. The two main reasons are (i) people with unlisted phone numbers are very likely to be the same kind of people that would seriously consider buying a burglar alarm, and (ii) the listing bias is more likely to be significant in a place like New York City. (People with unlisted phone numbers make up a much higher percentage of the population in a large city such as New York than in a small town or rural area. Interestingly enough, the largest percentage of unlisted phone numbers for any American city is in Las Vegas, Nevada.)

67. (a) Fridays, Saturdays, and Sundays make up 3/7 or about 43% of the week. It follows that proportionally, there are fewer fatalities due to automobile accidents on Friday, Saturday, and Sunday (42%) that there are on Monday through Thursday.

(b) On Saturday and Sunday there are fewer people commuting to work. The number of cars on the road and miles driven is significantly less on weekends. The number of accidents due to fatalities should be proportionally much less. The fact that it is 42% indicates that there are other factors involved and increased drinking is one possible explanation.

Chapter 14

WALKING

A.   Frequency Tables, Bar Graphs, and Pie Charts

1.

Score	10	50	60	70	80	100
Frequency	1	3	7	6	5	2

3. (a)

Grade	A	B	C	D	F
Frequency	7	6	7	3	1

(b)

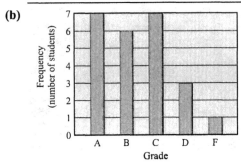

5. (a) $24 + 16 + 20 + 12 + 5 + 3 = 80$

(b) $\dfrac{24}{80} = 0.30$
30%

(c)

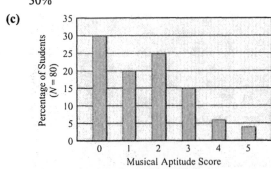

7.

Distance (miles) to school	0.0	0.5	1.0	1.5	2.0	2.5	3.0	5.0	8.5
Frequency	4	3	4	6	3	2	1	1	1

9. (a)

Class Interval	Very close	Close	Nearby	Not too far	Far
Frequency	7	10	5	1	2

(b)

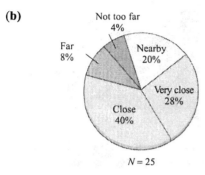

$N = 25$

Slice "Very close": $\dfrac{7}{25} \times 360° = 100.8°$; Slice "Close": $\dfrac{10}{25} \times 360° = 144°$;

Slice "Nearby": $\dfrac{5}{25} \times 360° = 72°$; Slice "Not too far": $\dfrac{1}{25} \times 360° = 14.4°$;

Slice "Far": $\dfrac{2}{25} \times 360° = 28.8°$.

11. (a) $N = 2 + 5 + 6 + 4 + 4 + 5 + 3 + 1 = 30$

(b) 0%

(c) $\dfrac{4 + 4 + 5 + 3 + 1}{30} \approx 0.5667$
approximately 56.67%

13. Accidents: $0.43023 \times 360° \approx 155°$
Homicide: $0.17297 \times 360° \approx 62°$
Suicide: $0.12982 \times 360° \approx 47°$
Cancer: $0.05236 \times 360° \approx 19°$
Heart Disease: $0.03208 \times 360° \approx 12°$
Other: $0.18254 \times 360° \approx 66°$

15. (a) $0.09 \times \$2,100,000,000,000 = \$189,000,000,000$ ($189 billion)

(b) $0.44 \times \$2,100,000,000,000 = \$924,000,000,000$ ($924 billion)

17.
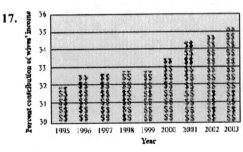
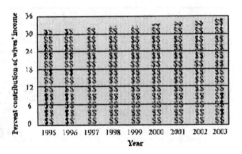

B.   Histograms

19. (a) $60 - 48 = 12$ ounces

(b) The third class interval: "more than 72 ounces and less than or equal to 84 ounces." Values that fall exactly on the boundary between two class intervals belong to the class interval to the left.

(c) $N = 15 + 24 + 41 + 67 + 119 + 184 + 142 + 26 + 5 + 2 = 625$

Frequency	15	24	41	67	119	184	142	26	5	2
Percent	2.4	3.8	6.6	10.7	19.0	29.4	22.7	4.2	0.8	0.3

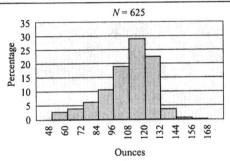

21. (a) $8 + 3 = 11$

(b) $20,000,000 ($20 million)

(c) Payrolls of $20 million, $40 million, $60 million, $80 million, or $100 million belong to the class interval on the right.

C.   Means and Medians

23. (a) average $= \dfrac{3-5+7+4+8+2+8-3-6}{9} = 2$

(b) The ordered data set is {-6, -5, -3, 2, 3, 4, 7, 8, 8}. The locator of the 50th percentile is L = (0.50)(9) = 4.5. Since L is not a whole number, the 50th percentile is located in the 5th position in the list. Hence, the median is 3.

(c) The average of the new data set is found by computing $\dfrac{(9\times 2)+2}{10}=2$ The new ordered data set is

$\{-6, -5, -3, 2, 2, 3, 4, 7, 8, 8\}$. The locator of the 50th percentile is L = (0.50)(10) = 5. Since L is a whole number, the 50th percentile is the average of the 5th and 6th numbers (2 and 3) in the ordered list. Hence, the median of this new data set is 2.5.

25. (a) average = $\dfrac{0+1+2+3+4+5+6+7+8+9}{10}=\dfrac{45}{10}=4.5$

Since the data set is already ordered, the locator of the median is L = (0.50)(10) = 5. Since L is a whole number, the median is the average of the 5th and 6th numbers (4 and 5) in the data set. Hence the median is 4.5.

(b) average = $\dfrac{1+2+3+4+5+6+7+8+9}{9}=\dfrac{45}{9}=5$

The locator of the median is L = (0.50)(9) = 4.5. Since L is not a whole number, the median is the 5th number in the data set. Hence the median is 5.

(c) average = $\dfrac{1+2+3+4+5+6+7+8+9+10}{10}=\dfrac{55}{10}=5.5$

The locator of the median is L = (0.50)(10) = 5. Since L is a whole number, the median is the average of the 5th and 6th numbers (5 and 6) in the data set. Hence the median is 5.5.

27. (a) average = $\dfrac{1+2+3+4+5+\ldots+98+99}{99}=\dfrac{\frac{99\times 100}{2}}{99}=50$

(b) The locator for the median is L = (0.50)(99) = 49.5. Hence, the median is in the 50th position. That is, the median is 50.

29. (a) average = $\dfrac{24\times 0+16\times 1+20\times 2+12\times 3+5\times 4+3\times 5}{24+16+20+12+5+3}=\dfrac{127}{80}=1.5875$

(b) The locator for the median is L = (0.50)(80) = 40. So, the median is the average of the 40th and 41st scores. The 40th score is 1, and the 41st score is 2. Thus, the median is 1.5.

31. (a) Since the number of quiz scores N is not given, we can compute the average as
$\dfrac{(0.07N)\times 4+(0.11N)\times 5+(0.19N)\times 6+(0.24N)\times 7+(0.39N)\times 8}{N}$
$=0.07\times 4+0.11\times 5+0.19\times 6+0.24\times 7+0.39\times 8$
$=6.77$

(As the calculation shows, the number of scores is not important. If one likes, they can assume that there were 100 scores.)

(b) Since 50% of the scores were at 7 or below, the median is 7.

D.   Percentiles and Quartiles

33. The ordered data set is {-6, -5, -3, 2, 3, 4, 7, 8, 8}.

(a) The locator of the 25th percentile is L = (0.25)(9) = 2.25. By rounding up, we find that the first quartile is the 3rd number in the ordered list. That is, $Q_1 = -3$.

(b) The locator of the 75th percentile is L = (0.75)(9) = 6.75. By rounding up, we find that the third quartile is the 7th number in the ordered list. That is, $Q_3 = 7$.

(c) The new ordered data set is {-6, -5, -3, 2, 2, 3, 4, 7, 8, 8}. The locator of the 25th percentile is L = (0.25)(10) = 2.5. Rounding up, we find that the first quartile is the 3rd number in the ordered list. That is, $Q_1 = -3$. The locator of the 75th percentile is L = (0.75)(10) = 7.5. Rounding up, we find that the third quartile is the 8th number in the ordered list. That is, $Q_3 = 7$.

35. (a) Since the data set is already ordered, the locator of the 75th percentile is given by L = (0.75)(100) = 75. Since this is a whole number, the 75th percentile is the average of the 75th and 76th numbers in the list. That is, the 75th percentile is 75.5.

The locator of the 90th percentile is given by L = (0.90)(100) = 90. Since this is a whole number, the 90th percentile is the average of the 90th and 91st numbers in the list. That is, the 90th percentile is 90.5.

(b) The locator of the 75th percentile is given by L = (0.75)(101) = 75.75. Since this not is a whole number, the 75th percentile is the 76th number in the list. That is, the 75th percentile is 75.

The locator of the 90th percentile is given by L = (0.90)(101) = 90.9. Since this is not a whole number, the 90th percentile is the 91st number in the list. That is, the 90th percentile is 90.

(c) The locator of the 75th percentile is given by L = (0.75)(99) = 74.25. Since this not is a whole number, the 75th percentile is the 75th number in the list. That is, the 75th percentile is 75.

The locator of the 90th percentile is given by L = (0.90)(99) = 89.1. Since this is not a whole number, the 90th percentile is the 90th number in the list. That is, the 90th percentile is 90.

(d) The locator of the 75th percentile is given by L = (0.75)(98) = 73.5. Since this not is a whole number, the 75th percentile is the 74th number in the list. That is, the 75th percentile is 74.

The locator of the 90th percentile is given by L = (0.90)(98) = 88.2. Since this is not a whole number, the 90th percentile is the 89th number in the list. That is, the 90th percentile is 89.

37. (a) The Cleansburg Fire Department consists of 2 + 7 + 6 + 9 + 15 + 12 + 9 + 9 + 6 + 4 = 79 firemen. The locator of the first quartile is thus given by L = (0.25)(79) = 19.75. So, the first quartile is the 20th number in the ordered data set. That is, $Q_1 = 29$.

(b) The locator of the third quartile is given by L = (0.75)(79) = 59.25. So, the third quartile is the 60th number in the ordered data set. That is, $Q_3 = 32$.

  (c) The locator of the 90th percentile is given by L = (0.90)(79) = 71.1. So, the 90th percentile is the 72nd number in the ordered data set or 37.

39. **(a)** The 709,504th score, $d_{709,504}$.
  The locator is given by $L = 709,503.5$.

  (b) The 354,752th score, $d_{354,752}$.
  The locator is given by $L = 354,751.75$.

  (c) The 1,064,256th score, $d_{1,064,256}$.
  The locator is given by $L = 1,064,255.25$.

E. **Box Plots and Five-Number Summaries**

41. **(a)** *Min* $= -6,\ Q_1 = -3, M = 3, Q_3 = 7,\ Max = 8$

  (b)

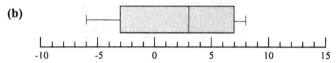

43. **(a)** *Min* $= 25,\ Q_1 = 29, M = 31, Q_3 = 32,\ Max = 39$

  (b)

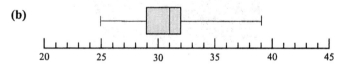

45. **(a)** Between $33,000 and $34,000 (corresponding the vertical line in the middle of the box)

  (b) $40,000

  (c) The vertical line indicating the median salary in the engineering box plot is to the right of the box in the agriculture box plot.

F. **Ranges and Interquartile Ranges**

47. **(a)** $8 - (-6) = 14$

  (b) From Exercise 33, $Q_1 = -3, Q_3 = 7$.
  $IQR = 7 - (-3) = 10$

49. **(a)** $156,000 - $115,000 = $41,000

  (b) At least 171 homes.

51. **(a)** Note that $1.5 \times IQR = 1.5 \times 3 = 4.5$. So any number bigger than $12 + 4.5 = 16.5$ is an outlier.

  (b) Any number smaller than $9 - 4.5 = 4.5$ is also an outlier.

(c) Since 1 is the only number smaller than 4.5 and 24 is the only number bigger than 16.5, the numbers 1 and 24 are the only outliers in this data set.

53. The *IQR* for the 2001 SAT math scores is $590 - 440 = 150$. Since $1.5 \times IQR = 1.5 \times 150 = 225$, an outlier is any score bigger than $590 + 225 = 815$ or any score smaller than $440 - 225 = 215$. Since the maximum score on the SAT is 800 and the minimum SAT score is 200, the only outliers are scores less than 215. That is, only SAT math scores of 200 or 210 would be considered outliers.

G.   Standard Deviations

55. (a) $A = 5$, so $x - A = 0$ for every number x in the data set. The standard deviation is 0.

(b) $A = \dfrac{0 + 5 + 5 + 10}{4} = 5$

x	$x - 5$	$(x-5)^2$
0	−5	25
5	0	0
5	0	0
10	5	25
		50

Standard deviation $= \sqrt{\dfrac{50}{4}} = \dfrac{5\sqrt{2}}{2} \approx 3.5$

(c) $A = \dfrac{-5 + 0 + 0 + 25}{4} = 5$

x	$x - 5$	$(x-5)^2$
−5	−10	100
0	−5	25
0	−5	25
25	20	400
		550

Standard deviation $= \sqrt{\dfrac{550}{4}} \approx 11.7$

57. (a) $A = \dfrac{0 + 1 + 2 + 3 + 4 + 5 + 6 + 7 + 8 + 9}{10} = 4.5$

x	$x - 4.5$	$(x-4.5)^2$
0	−4.5	20.25
1	−3.5	12.25
2	−2.5	6.25
3	−1.5	2.25
4	−0.5	0.25
5	0.5	0.25
6	1.5	2.25
7	2.5	6.25
8	3.5	12.25
9	4.5	20.25
		82.5

$$\text{Standard deviation} = \sqrt{\frac{82.5}{10}} \approx 2.87$$

(b) $A = \dfrac{1+2+3+4+5+6+7+8+9+10}{10} = 5.5$

x	$x - 5.5$	$(x-5.5)^2$
1	−4.5	20.25
2	−3.5	12.25
3	−2.5	6.25
4	−1.5	2.25
5	−0.5	0.25
6	0.5	0.25
7	1.5	2.25
8	2.5	6.25
9	3.5	12.25
10	4.5	20.25
		82.5

Standard deviation $= \sqrt{\dfrac{82.5}{10}} \approx 2.87$

Note that each data point is simply located 1 unit to the right of that data set given in (a). So, the spread of the data set has not changed.

(c) $A = \dfrac{6+7+8+9+10+11+12+13+14+15}{10} = 10.5$

x	$x - 10.5$	$(x-10.5)^2$
6	−4.5	20.25
7	−3.5	12.25
8	−2.5	6.25
9	−1.5	2.25
10	−0.5	0.25
11	0.5	0.25
12	1.5	2.25
13	2.5	6.25
14	3.5	12.25
15	4.5	20.25
		82.5

Standard deviation $= \sqrt{\dfrac{82.5}{10}} \approx 2.87$

Note again that each data point is 6 units to the right of where it appeared in (a). So, the mean has changed (it is 6 units bigger), but the spread of the data has not. So, the standard deviation is the same as in (a).

(d) $A = \dfrac{5+15+25+35+45+55+65+75+85+95}{10} = 50$

x	$x - 50$	$(x-50)^2$
5	−45	2025
15	−35	1225
25	−25	625
35	−15	225
45	−5	25
55	5	25

65	15	225
75	25	625
85	35	1225
95	45	2025
		8250

Standard deviation = $\sqrt{\dfrac{8250}{10}} \approx 28.7$

Note that each data point is 10 times farther away from the mean as in (a). So the spread (standard deviation) is 10 times larger.

H.   Miscellaneous

59. 10 (Frequency 16)

61. 4, 5, and 8 (Frequency 5)

63. Caucasian (Has largest percent.)

JOGGING

65. Let x = Mike's score on the next exam.

$\dfrac{5 \times 88 + x}{6} = 90$

$5 \times 88 + x = 540$

$x = 100$

67. Ramon gets 85 out of 100 on each of the first four exams and 60 out of 100 on the fifth exam. Josh gets 80 out of 100 on all 5 of the exams.

69. **(a)** {1, 1, 1, 1, 6, 6, 6, 6, 6, 6} Average = 4; Median = 6.

   (b) {1, 1, 1, 1, 1, 1, 6, 6, 6, 6} Average = 3; Median =1.

   (c) {1, 1, 6, 6, 6, 6, 6, 6, 6, 6} Average = 5; $Q_1 = 6$.

   (d) {1, 1, 1, 1, 1, 1, 1, 1, 6, 6} Average = 2; $Q_3 = 1$.

71. **(a)** The five-number summary for the original scores was *Min* = 1, $Q_1 = 9, M = 11, Q_3 = 12,$ and *Max* = 24. When 2 points are added to each test score, the five-number summary will also have 2 points added to each of its numbers (i.e., *Min* = 3, $Q_1 = 11, M = 13, Q_3 = 14,$ and *Max* = 26).

   (b) When 10% is added to each score (i.e., each score is multiplied by 1.1) then each number in the five-number summary will also be multiplied by 1.1 (i.e., *Min* = 1.1, $Q_1 = 9.9, M = 12.1, Q_3 = 13.2,$ and *Max* = 26.4).

73. (a) 4

$$\frac{\text{Column area over interval } 30-35}{\text{Column area over interval } 20-30} = \frac{5 \times h}{10 \times 1} = \frac{50\%}{25\%} \text{ and so } 5h = 20, \ h = 4.$$

(b) 0.4

$$\frac{\text{Column area over interval } 35-45}{\text{Column area over interval } 20-30} = \frac{10 \times h}{10 \times 1} = \frac{10\%}{25\%} \text{ and so } h = 0.4.$$

(c) 0.4

$$\frac{\text{Column area over interval } 45-60}{\text{Column area over interval } 20-30} = \frac{15 \times h}{10 \times 1} = \frac{15\%}{25\%} \text{ and so } h = 0.4.$$

75. (a) Male: 10%, Female: 20%

(b) Male: 80%, Female: 90%

(c) The figures for both schools were combined. A total of 820 males were admitted out of a total of 1200 that applied–an admission rate for males of approximately 68.3%. Similarly, a total of 460 females were admitted out of a total of 900 that applied–an admission rate for females of approximately 51.1%.

(d) In this example, females have a higher percentage $\left(\frac{100}{500} = 20\% \right)$ than males $\left(\frac{20}{200} = 10\% \right)$ for admissions to the School of Architecture and also a higher percentage $\left(\frac{360}{400} = 90\% \right)$ than males $\left(\frac{800}{1000} = 80\% \right)$ for the School of Engineering. When the numbers are combined, however, females have a lower percentage $\left(\frac{100+360}{500+400} \approx 51.1\% \right)$ than males $\left(\frac{20+800}{200+1000} \approx 68.3\% \right)$ in total admissions. The reason that this apparent paradox can occur is purely a matter of arithmetic: Just because $\frac{a_1}{a_2} > \frac{b_1}{b_2}$ and $\frac{c_1}{c_2} > \frac{d_1}{d_2}$ it does not necessarily follow that $\frac{a_1 + c_2}{a_2 + c_2} > \frac{b_1 + d_1}{b_2 + d_2}$. The majority of males applied to the engineering school, which has a much higher acceptance rate than the School of Architecture. This, combined with the fact that more females applied to the School of Architecture than the School of Engineering, is why, overall, the percentage of males admitted is higher than the percentage of females admitted.

77. The relative sizes of the numbers are not changed by adding a constant c to every number. Thus (assuming the original data set is sorted) if the median of the data set is $M = x_k$, then the median of the new data set will be $x_k + c = M + c$. If the median of the original set is $M = \frac{x_k + x_{k+1}}{2}$, then the median of the new data set will be $\frac{(x_k + c) + (x_{k+1} + c)}{2} = \frac{x_k + x_{k+1}}{2} + \frac{2c}{2} = M + c$. In either case, we see that the median of the new data set is $M + c$.

79. (a)

Chapter	1	2	3	4	5	6	7	8	9	10	11	12	13	14	15	16
Exercises	80	78	84	67	73	75	76	75	78	80	80	75	68	89	86	81

$$\text{average} = \frac{1245}{16} = 77.8125 \ \text{exercises}$$

(b) 78

The sorted data set: $\{67, 68, 73, 75, 75, 75, 76, 78, 78, 80, 80, 80, 81, 84, 86, 89\}$

The median is the average of the 8th and 9th elements of this sorted data set. That is, the average of 78 and 78.

(c) $Min = 67, \ Q_1 = 75, \ M = 78, \ Q_3 = 80.5, \ Max = 89$

(d)

x	$x - 77.8125$	$(x - 77.8125)^2$
80	2.1875	4.78515625
78	0.1875	0.03515625
84	6.1875	38.28515625
67	-10.8125	116.9101563
73	-4.8125	23.16015625
75	-2.8125	7.91015625
76	-1.8125	3.28515625
75	-2.8125	7.91015625
78	0.1875	0.03515625
80	2.1875	4.78515625
80	2.1875	4.78515625
75	-2.8125	7.91015625
68	-9.8125	96.28515625
89	11.1875	125.1601563
86	8.1875	67.03515625
81	3.1875	10.16015625
		518.4375

$$\text{standard deviation} = \sqrt{\frac{518.4375}{16}} \approx 5.69 \ \text{exercises}$$

Chapter 15

WALKING

A.   Random Experiments and Sample Spaces

1. **(a)** {*HHH, HHT, HTH, THH, TTH, THT, HTT, TTT*}

    (b) {0, 1, 2, 3}

    (c) {0, 1, 2, 3}

3. **(a)** {*ABCD, ABDC, ACBD, ACDB, ADBC, ADCB, BACD, BADC, BCAD, BCDA, BDAC, BDCA, CABD, CADB, CBAD, CBDA, CDAB, CDBA, DABC, DACB, DBAC, DBCA, DCAB, DCBA*}

    (b) $N = 4 \times 3 \times 2 \times 1 = 24$ (4 choices for first name chosen, 3 choices for second name chosen, etc.)

5. Answers will vary. A typical outcome is a string of 10 letters each of which can be either an *H* or a *T*. An answer like {*HHHHHHHHHH*, …,*TTTTTTTTTT*} is not sufficiently descriptive. An answer like{… *HTTHHHTHTH*, …, *TTHTHHTTHT*, …, *HHHTHTTTHHT*, …} is better. An answer like $\{(X_1 X_2 X_3 X_4 X_5 X_6 X_7 X_8 X_9 X_{10}) :$ each X_i is either *H* or *T*} is best. Note: This sample space consists of $N = 2^{10} = 1024$ outcomes (2 possible outcomes at each of 10 stages of the experiment).

7. Answers will vary. An outcome is an ordered sequence of four numbers, each of which is an integer between 1 and 6 inclusive. The best answer would be something like $\{(n_1, n_2, n_3, n_4) :$ each n_i is 1, 2, 3, 4, 5, or 6}. An answer such as {(1,1,1,1), …, (1,1,1,6), …,(1,2,3,4),…, (3,2,6,2), …, (6,6,6,6)} showing a few typical outcomes is possible, but not as good. An answer like {(1,1,1,1),…, (2,2,2,2), …, (6,6,6,6)} is not descriptive enough. Note: This sample space consists of $N = 6^4 = 1296$ outcomes (6 possible outcomes at each of 4 stages of the experiment).

B.   The Multiplication Rule.

9. **(a)** $9 \times 26^3 \times 10^3 = 159,184,000$

    (b) $1 \times 26^3 \times 10^2 \times 1 = 1,757,600$

    (c) $9 \times 26 \times 25 \times 24 \times 9 \times 8 \times 7 = 70,761,600$

11. **(a)** $52^4 \times 10 = 73,116,160$

    (b) $52^3 \times 10 = 1,406,080$

    (c) $50 \times 52^3 \times 10 = 70,304,000$

    (d) $50^4 \times 10 = 62,500,000$

13. **(a)** $8! = 40,320$

(b) $40,320 - 1 = 40,319$ (there is only one way in which all of the books are in order)

15. (a) $35 \times 34 \times 33 = 39,270$

(b) $15 \times 34 \times 33 = 16,830$

(c) The total number of all-girl committees is $15 \times 14 \times 13 = 2730$.
The total number of all-boy committees is $20 \times 19 \times 18 = 6840$.
The remaining $35 \times 34 \times 33 - (15 \times 14 \times 13 + 20 \times 19 \times 18) = 39,270 - (2730 + 6840) = 29,700$ committees are mixed.

17. (a) $9 \times 10^5 \times 5 = 4,500,000$
There are 9 choices for the first digit (1-9), 10 choices for the next 5 digits (0-9), and 5 digits for the last digit (0, 2, 4, 6, 8).

(b) $9 \times 10^5 \times 2 = 1,800,000$
There are 2 choices for the last digit (either a 0 or a 5).

(c) $9 \times 10^4 \times 4 = 360,000$
The last 2 digits must be 00, 25, 50 or 75.

C.   Permutations and Combinations

19. (a) $_{10}P_2 = 10 \times 9 = 90$

(b) $_{10}C_2 = \dfrac{10 \times 9}{2 \times 1} = \dfrac{90}{2} = 45$

(c) $_{10}P_3 = 10 \times 9 \times 8 = 720$

(d) $_{10}C_3 = \dfrac{720}{3!} = \dfrac{720}{6} = 120$

21. (a) $_{10}C_9 = \dfrac{10!}{(10-9)!9!} = \dfrac{10!}{1!9!} = 10$

(b) $_{10}C_8 = \dfrac{10!}{(10-8)!8!} = \dfrac{10!}{2!8!} = \dfrac{10 \times 9}{2 \times 1} = 45$

(c) $_{100}C_{99} = \dfrac{100!}{(100-99)!99!} = \dfrac{100!}{1!99!} = 100$

(d) $_{100}C_{98} = \dfrac{100!}{(100-98)!98!} = \dfrac{100!}{2!98!} = \dfrac{100 \times 99}{2 \times 1} = 4950$

23. (a) $_{20}C_2 = \dfrac{20!}{(20-2)!2!} = \dfrac{20!}{18!2!} = \dfrac{20 \times 19}{2 \times 1} = 190$

(b) $_{20}C_{18} = \dfrac{20!}{(20-18)!18!} = \dfrac{20!}{2!18!} = \dfrac{20 \times 19}{2 \times 1} = 190$

(c) $_{20}C_3 = \dfrac{20!}{(20-3)!3!} = \dfrac{20!}{17!3!} = \dfrac{20 \times 19 \times 18}{3 \times 2 \times 1} = 1140$

(d) $_{20}C_{17} = \dfrac{20!}{(20-17)!17!} = \dfrac{20!}{3!17!} = \dfrac{20 \times 19 \times 18}{3 \times 2 \times 1} = 1140$

25. (a) $_3C_0 + _3C_1 + _3C_2 + _3C_3 = \dfrac{3!}{(3-0)!0!} + \dfrac{3!}{(3-1)!1!} + \dfrac{3!}{(3-2)!2!} + \dfrac{3!}{(3-3)!3!}$

$$= 1 + 3 + 3 + 1$$
$$= 8$$

(b) $_4C_0 + _4C_1 + _4C_2 + _4C_3 + _4C_4$

$$= \dfrac{4!}{(4-0)!0!} + \dfrac{4!}{(4-1)!1!} + \dfrac{4!}{(4-2)!2!} + \dfrac{4!}{(4-3)!3!} + \dfrac{4!}{(4-4)!4!}$$
$$= 1 + 4 + 6 + 4 + 1$$
$$= 16$$

(c) $1 + 5 + 10 + 10 + 5 + 1 = 32$

(d) The pattern in (a)-(c) is that each of these sums is a power of two. In (a), the sum was $2^3 = 8$. In (b), the sum was $2^4 = 16$. In (c), the sum was $2^5 = 32$. It should then be that the last sum is $2^{10} = 1,024$. (One should also note that values found in the rows of Pascal's triangle are being added.)

27. First, note that $_{150}P_{50} = \dfrac{150!}{100!} = \dfrac{150 \times 149 \times 148 \times \cdots \times 3 \times 2 \times 1}{100 \times 99 \times 98 \times \cdots \times 3 \times 2 \times 1}$ and

$_{150}P_{51} = \dfrac{150!}{99!} = \dfrac{150 \times 149 \times 148 \times \cdots \times 3 \times 2 \times 1}{99 \times 98 \times 97 \times \cdots \times 3 \times 2 \times 1}$. It follows that $_{150}P_{51} = 100 \times _{150}P_{50}$. That is, $_{150}P_{51}$ is 100 times bigger than $_{150}P_{50}$. Thus, $_{150}P_{51} = 6.12 \times 10^{104} \times 100 = 6.12 \times 10^{106}$.

29. (a) $_{20}P_{10} = 670,442,572,800 \approx 6.70 \times 10^{11}$

(b) $_{52}C_{20} = 125,994,627,894,135 \approx 1.26 \times 10^{14}$

(c) $_{52}C_{32} = 125,994,627,894,135 \approx 1.26 \times 10^{14}$

    (d) $_{53}C_{20} = 202,355,008,436,035 \approx 2.02 \times 10^{14}$

31. (a) $_{15}P_4$

    (b) $_{15}C_4$

33. (a) $_{20}C_2$

    (b) $_{20}P_2$

    (c) $_{20}C_5$

D.   General Probability Spaces

35. (a) $\Pr(o_1) + \Pr(o_2) + \Pr(o_3) + \Pr(o_4) + \Pr(o_5) = 1$
$$0.22 + 0.24 + 3\Pr(o_3) = 1$$
$$\Pr(o_3) = 0.18$$

    (b) $\Pr(o_1) + \Pr(o_2) + \Pr(o_3) + \Pr(o_4) + \Pr(o_5) = 1$
$$0.22 + 0.24 + 2\Pr(o_3) = 1$$
$$\Pr(o_3) = 0.27$$

    (c) $\Pr(o_1) + \Pr(o_2) + \Pr(o_3) + \Pr(o_4) + \Pr(o_5) = 1$
$$0.22 + 0.24 + 0.27 + \Pr(o_4) + 0.1 = 1$$
$$\Pr(o_4) = 0.17$$
The probability assignment is $\Pr(o_1) = 0.22$, $\Pr(o_2) = 0.24$, $\Pr(o_3) = 0.27$, $\Pr(o_4) = 0.17$, $\Pr(o_5) = 0.1$.

37. $S = \{P_1, P_2, P_3, P_4, P_5, P_6, P_7\}$
$$\Pr(P_1) + \Pr(P_2) + \Pr(P_3) + \Pr(P_4) + \Pr(P_5) + \Pr(P_6) + \Pr(P_7) = 1$$
$$8\Pr(P_2) = 1$$
$$\Pr(P_2) = \frac{1}{8}$$
$$\Pr(P_1) = 2\Pr(P_2) = 2\left(\frac{1}{8}\right) = \frac{2}{8}$$
The probability assignment is $\Pr(P_1) = \dfrac{2}{8}$, $\Pr(P_2) = \Pr(P_3) = \Pr(P_4) = \Pr(P_5) = \Pr(P_6) = \Pr(P_7) = \dfrac{1}{8}$.

39. (a) $S = \{\text{red, blue, white, green, yellow}\}$

(b) $\Pr(\text{blue}) = \Pr(\text{white}) = \dfrac{72°}{360°} = 0.2$

$\Pr(\text{green}) = \Pr(\text{yellow}) = \dfrac{54°}{360°} = 0.15$

The probability assignment is $\Pr(\text{red}) = 0.3$, $\Pr(\text{blue}) = \Pr(\text{white}) = 0.2$, $\Pr(\text{green}) = \Pr(\text{yellow}) = 0.15$.

E.   Events

41. (a) $E_1 = \{HHT, HTH, THH\}$

(b) $E_2 = \{HHH, TTT\}$

(c) $E_3 = \{\,\}$

(d) $E_4 = \{TTH, TTT\}$

43. (a) $E_1 = \{(1,1),(2,2),(3,3),(4,4),(5,5),(6,6)\}$

(b) $E_2 = \{(1,1),(1,2),(2,1)\}$

(c) $E_3 = \{(1,6),(2,5),(3,4),(4,3),(5,2),(6,1),(5,6),(6,5)\}$

45. (a) $E_1 = \{HHHHHHHHHH\}$

(b) $E_2 = \{HHHHHHHHHT, HHHHHHHHTH, HHHHHHHTHH, HHHHHHTHHH, HHHHHTHHHH,$
$HHHHTHHHHH, HHHTHHHHHH, HHTHHHHHHH, HTHHHHHHHH, THHHHHHHHH\}$

(c) $E_3 = \{\,\}$

F.   Equiprobable Spaces

47. (a) $\Pr(E_1) = \dfrac{3}{8} = 0.375$

(b) $\Pr(E_2) = \dfrac{2}{8} = 0.25$

(c) $\Pr(E_3) = 0$

(d) $\Pr(E_4) = \dfrac{2}{8} = 0.25$

49. (a) $T_6 = \{(1,5),(2,4),(3,3),(4,2),(5,1)\}$ and $T_8 = \{(2,6),(3,5),(4,4),(5,3),(6,2)\}$

So, $\Pr(T_6) = \dfrac{5}{36}$ and $\Pr(T_8) = \dfrac{5}{36}$.

(b) $T_5 = \{(1,4),(2,3),(3,2),(4,1)\}$ and $T_9 = \{(3,6),(4,5),(5,4),(6,3)\}$

So, $\Pr(T_5) = \dfrac{4}{36} = \dfrac{1}{9}$ and $\Pr(T_9) = \dfrac{4}{36} = \dfrac{1}{9}$.

(c) $\Pr(E_1) = \dfrac{6}{36} = \dfrac{1}{6} \approx 0.167$ (see Exercise 43(a))

(d) $\Pr(E_2) = \dfrac{3}{36} = \dfrac{1}{12} \approx 0.083$ (see Exercise 43(b))

(e) $\Pr(E_3) = \dfrac{8}{36} = \dfrac{2}{9} \approx 0.222$ (see Exercise 43(c))

51. (a) $\Pr(E_1) = \dfrac{1}{1024} \approx 0.001$

(b) $\Pr(E_2) = \dfrac{10}{1024} = \dfrac{5}{512} \approx 0.01$

(c) $\Pr(E_3) = 0$

53. (a) $\Pr(\text{"roll 8"}) = \dfrac{5}{36} \approx 0.139$

(b) $\Pr(\text{"not roll 8"}) = 1 - \dfrac{5}{36} = \dfrac{31}{36} \approx 0.861$

(c) $\Pr(\text{"roll 8 or 9"}) = \dfrac{9}{36} = \dfrac{1}{4} = 0.25$

(d) $\Pr(\text{"roll 8 or more"}) = \dfrac{15}{36} \approx 0.417$

55. The total number of outcomes in this random experiment is $2^{10} = 1024$.

(a) There is only one way to get all ten correct. So, $\Pr(\text{getting 10 points}) = \dfrac{1}{1024}$.

(b) There is only one way to get all ten incorrect (and hence a score of -5). So,

$$\text{Pr(getting -5 points)} = \frac{1}{1024}.$$

(c) In order to get 8.5 points, the student must get exactly 9 correct answers and 1 incorrect answer. There are $_{10}C_1 = 10$ ways to select which answer would be answered incorrectly. So,

$$\text{Pr(getting 8.5 points)} = \frac{10}{1024} = \frac{5}{512}.$$

(d) In order to get 8 or more points, the student must get at least 9 correct answers (if they only get 8 correct answers, they lose a point for guessing two incorrect answers and score 7 points). So,

$$\text{Pr(getting 8 or more points)} = \text{Pr(getting 8.5 points)} + \text{Pr(10 points)} = \frac{10}{1024} + \frac{1}{1024} = \frac{11}{1024}.$$

(e) If the student gets 6 answers correct, they score $6 - 4 \times 0.5 = 4$ points. If the student gets 7 answers correct, they score $7 - 3 \times 0.5 = 5.5$ points. So, there is no chance of getting exactly 5 points. Hence, $\text{Pr(getting 5 points)} = 0$.

(f) If the student gets 8 answers correct, they score $8 - 2 \times 0.5 = 7$ points. There are $_{10}C_2 = 45$ ways to select which 8 answers would be answered correctly. So, $\text{Pr(getting 7 points)} = \dfrac{45}{1024}$.

In order to get 7 or more points, the student needs to answer at least 8 answers correctly.
$$\text{Pr(getting 7 or more points)} = \text{Pr(getting 7 points)} + \text{Pr(getting 8.5 points)} + \text{Pr(10 points)}$$

$$= \frac{45}{1024} + \frac{10}{1024} + \frac{1}{1024}$$
$$= \frac{56}{1024}$$
$$= \frac{7}{128}$$

57. (a) There are $_{15}C_4 = 1365$ ways to choose four delegates. If Alice is selected, there are $_{14}C_3 = 364$ ways to choose the other three delegates.

$$\text{Pr(Alice selected)} = \frac{364}{1365} = \frac{4}{15} \approx 0.267$$

(b) $\text{Pr(Alice not selected)} = 1 - \dfrac{364}{1365} = \dfrac{1001}{1365} = \dfrac{11}{15} \approx 0.733$

(c) There are $_{15}C_4 = 1,365$ ways to select four members, but only one way to select Alice, Bert, Cathy, and Dale.

$$\text{Pr(Alice, Bert, Cathy, and Dale selected)} = \frac{1}{_{15}C_4} = \frac{1}{1,365} \approx 0.0007326$$

G.   Odds

59. (a) $a = 4, b = 7, b - a = 7 - 4 = 3$
The odds in favor of *E* are 4 to 3.

(b) $a = 6, b = 10, b - a = 10 - 6 = 4$
The odds in favor of *E* are 6 to 4, or 3 to 2.

61. (a) $\text{Pr}(E) = \dfrac{3}{3+5} = \dfrac{3}{8}$

(b) $\text{Pr}(E) = 1 - \dfrac{8}{8+15} = 1 - \dfrac{8}{23} = \dfrac{15}{23}$

(c) $\text{Pr}(E) = \dfrac{1}{1+1} = \dfrac{1}{2}$

JOGGING

63. There are 35 ways to select a chair and 34 ways to select a secretary. From the remaining 33 members of the ski club, there are $_{33}C_3 = 5456$ ways to select three at-large members. This makes a total of $35 \times 34 \times 5456 = 6,492,640$ ways to select a planning committee.

65. (a) The event that X wins in 5 games can be described by
{*YXXXX, XYXXX, XXYXX, XXXYX*}.

(b) The event that the series is over in 5 games can be described by
{*YXXXX, XYXXX, XXYXX, XXXYX, XYYYY, YXYYY, YYXYY, YYYXY*}.

(c)

	X wins	*Y* wins
6 game series	*YYXXXX, YXYXXX, YXXYXX,* *YXXXYX, XYYXXX, XYXYXX,* *XYXXYX, XXYYXX, XXYXYX,* *XXXYYX*	*XXYYYY, XYXYYY, XYYXYY,* *XYYYXY, YXXYYY, YXYXYY,* *YXYYXY, YYXXYY, YYXYXY,* *YYYXXY*

The event "the series is over in game 6" can be described by {*YYXXXX, YXYXXX, YXXYXX, YXXXYX, XYYXXX, XYXYXX, XYXXYX, XXYYXX, XXYXYX, XXXYYX, XXYYYY, XYXYYY, XYYXYY, XYYYXY, YXXYYY, YXYXYY, YXYYXY, YYXXYY, YYXYXY, YYYXXY*}

67. (a) $10! = 3,628,800$

(b) A circle of 10 people can be broken to form a line in 10 different ways.
So there are $\dfrac{3,628,800}{10} = 362,880$ ways to form a circle.

(c) There are 2 choices as to which sex will start the line. Then, there are 5! ways to order the boys in the line and 5! ways to order the girls in the line. So, there are $2 \times 5! \times 5! = 28,800$ ways to form such a line.

(d) $\dfrac{2 \times 5! \times 5!}{10} = 2,880$ ways

69. Suppose, for the moment, that the order the teams (but not their members) were selected mattered. There are $\left({}_{15}C_5 \right) \cdot \left({}_{10}C_5 \right) \cdot \left({}_{5}C_5 \right) = 756,756$ ways to select the teams. Since the order that the teams are selected does not matter, and there are $3! = 6$ ways to rearrange the teams, there are actually $\dfrac{756,756}{6} = 126,126$ ways to select the teams.

As an alternative method of solution, think in the following way: Start with an ordered list of the study group. The first person on the list must be in *some* group. Then, select 4 of the remaining 14 members to be in that group (${}_{14}C_4 = 1001$ ways to do this). The first person on the list of the remaining members must be in some group (one of the two groups left to be formed). Then, select 4 of the remaining 9 members to be in that group (${}_{9}C_4 = 126$ ways to do this). Finally, the remaining 5 people on the list must belong to the study group not yet formed. So, there are $1001 \times 126 = 126,126$ ways to form the three groups.

71. The total number of possible outcomes is $2^{20} = 1,048,576$.

(a) Choose the positions of the *H*'s: ${}_{20}C_{10} = 184,756$.
$$\Pr(10\ H\text{'s and 10 }T\text{'s}) = \frac{184,756}{1,048,576} = \frac{46,189}{262,144} \approx 0.176$$

(b) Choose the positions of the *H*'s: ${}_{20}C_3 = 1140$.
$$\Pr(3\ H\text{'s and 17 }T\text{'s}) = \frac{1140}{1,048,576} = \frac{285}{262,144} \approx 0.001$$

(c) $\Pr(3 \text{ or more } H\text{'s}) = 1 - \Pr(0\ H\text{'s}) - \Pr(1\ H\text{'s}) - \Pr(2\ H\text{'s})$
$$= 1 - \frac{{}_{20}C_0}{2^{20}} - \frac{{}_{20}C_1}{2^{20}} - \frac{{}_{20}C_2}{2^{20}}$$
$$= 1 - \frac{1}{2^{20}} - \frac{20}{2^{20}} - \frac{190}{2^{20}}$$
$$= 1 - \frac{211}{1,048,576}$$
$$= \frac{1,048,365}{1,048,576} \approx 0.9998$$

73. The total number of 5-card draw poker hands is 2,598,960. The number of hands with all 5 cards the same color is $\dfrac{52 \times 25 \times 24 \times 23 \times 22}{5!} = 131,560.$

Pr(all 5 same color) $= \dfrac{131,560}{2,598,960} = \dfrac{253}{4,998} \approx 0.05$

75. The total number of 5-card draw poker hands is 2,598,960. The number of ways to get 10, J, Q, K, A of any suit (including all the same suit) is $\dfrac{20 \times 16 \times 12 \times 8 \times 4}{5!} = 1024.$

There are 4 ways for these cards to all be the same suit. So there are $1024 - 4 = 1020$ ways to get an ace-high straight.

Pr(ace-high straight) $= \dfrac{1020}{2,598,960} = \dfrac{1}{2548} \approx 0.00039$

77. Pr(win) $= 1 - $ Pr(never roll a 7)

$$= 1 - \left(\frac{30}{36}\right)^5$$

$$= 1 - \left(\frac{5}{6}\right)^5 = \frac{4,651}{7,776} \approx 0.6$$

79. (a) $(1 - 0.02)^{12} = (0.98)^{12} \approx 0.78.$ We are assuming that the events are independent.

(b) Pr(at most 1 defective) = Pr (none defective) + Pr(1 defective)

$$= (0.98)^{12} + 12(0.98)^{11}(0.02)$$

$$\approx 0.98$$

Mini-Excursion 4

WALKING

A.   Weighted Averages

1. Using the fact that 90/120 = 75% and 144/180 = 80%, Paul's score is
$0.15 \times 77\% + 0.15 \times 83\% + 0.15 \times 91\% + 0.1 \times 75\% + 0.25 \times 87\% + 0.2 \times 80\% = 82.9\%$

3. 100% - 7% - 22% - 24% - 23% - 19% = 5% are 19 years old. So, the average age at Thomas Jefferson HS is $0.07 \times 14 + 0.22 \times 15 + 0.24 \times 16 + 0.23 \times 17 + 0.19 \times 18 + 0.05 \times 19 = 16.4$.

B.   Expected Values

5. The expected value of this random variable is $E = \frac{1}{5} \times 5 + \frac{2}{5} \times 10 + \frac{2}{5} \times 15 = 11$.

7. (a)

Outcome	$1	$5	$10	$20	$100
Probability	1/2	1/4	1/8	1/10	1/40

(b) $E = \frac{1}{2} \times \$1 + \frac{1}{4} \times \$5 + \frac{1}{8} \times \$10 + \frac{1}{10} \times \$20 + \frac{1}{40} \times \$100 = \$7.50$

(c) $7.50

9. (a)

Outcome	0	1	2	3
Probability	1/8	3/8	3/8	1/8

(b) $E = \frac{1}{8} \times 0 + \frac{3}{8} \times 1 + \frac{3}{8} \times 2 + \frac{1}{8} \times 3 = 1.5$ heads

11. (a)

Outcome (Profit)	$1 (red)	$(-1) (black)	$(-1) (green)
Probability	18/38	18/38	2/38

$E = \frac{18}{38} \times \$1 + \frac{18}{38} \times \$(-1) + \frac{2}{38} \times \$(-1) = \$ -\frac{1}{19} \approx \$ -0.05$

(b) $E = \frac{18}{38} \times \$N + \frac{18}{38} \times \$(-N) + \frac{2}{38} \times \$(-N) = \$ -\frac{1}{19} N \approx \$ -0.05N$. So, for every $100 bet on red, you should expect to lose about $5 ($5.26 to be more precise). For every $1,000,000 bet on red, you should expect to lose about $52,631.

13. (a) $E = \frac{1}{6} \times \$1 + \frac{1}{6} \times \$(-2) + \frac{1}{6} \times \$3 + \frac{1}{6} \times \$(-4) + \frac{1}{6} \times \$5 + \frac{1}{6} \times \$(-6) = \$ - 0.50$

(b) Pay $0.50 to play a game in which you roll a single die. If an odd number comes up, you have to pay the amount of your roll ($1, $3, or $5). If an even number (2, 4, or 6) comes up, you win the amount of your roll.

C. Miscellaneous

15.

Outcome (Benefit to Joe)	$320	$(-80)
Probability	24%	76%

$E = 0.24 \times \$320 + 0.76 \times \$(-80) = \$16$

Joe should take this risk since his expected payoff is greater than 0 ($16). That is, if he made this transaction on thousands of plasma TVs, in the long run he could expect to save $16 for each warranty he purchases.

17.

Payoff (to insurer)	$P	$(P-500)	$(P-1,500)	$(P-4,000)
Probability	50%	35%	12%	3%

In order to make an average profit of $50 per policy, we solve

$E = 0.5 \times \$P + 0.35 \times \$(P - 500) + 0.12 \times \$(P - 1,500) + 0.03 \times \$(P - 4,000) = \$50$ for P. Doing this gives $P = \$525$. That is, the insurance company should charge $525 per policy.

19. (a) There are three ways to select which of the three dice will not land as a 4. There are five numbers that this die can land on (1, 2, 3, 5, or 6). The multiplication rule tells us that there are $3 \times 5 = 15$ ways to select an outcome in which exactly two 4's are rolled. Since there are $6^3 = 216$ (again by the multiplication rule) possible outcomes in this random experiment, the probability of such an outcome is 15/216.

(b) There are three ways to select which of the three dice will land as a 4. There are five numbers that each other die can land on (1, 2, 3, 5, or 6). The multiplication rule tells us that there are $3 \times 5 \times 5 = 75$ ways to select an outcome in which exactly one 4 is rolled. Hence, the probability of such an outcome is 75/216.

(c) There are five numbers that each die can land on (1, 2, 3, 5, or 6). So, there are $5 \times 5 \times 5 = 125$ ways to select an outcome in which no 4 is rolled. So, the probability of not rolling any 4's is 125/216.

21.

Outcome (Winnings)	$29,999,999	$(-1)
Probability	$\dfrac{1}{{}_{47}C_5 \times {}_{27}C_1}$	$1 - \dfrac{1}{{}_{47}C_5 \times {}_{27}C_1}$

$$E = \frac{1}{{}_{47}C_5 \times {}_{27}C_1} \times \$29{,}999{,}999 + \left(1 - \frac{1}{{}_{47}C_5 \times {}_{27}C_1}\right) \times \$(-1) =$$

$$\frac{1}{41{,}416{,}353} \times \$29{,}999{,}999 + \frac{41{,}415{,}352}{41{,}416{,}353} \times \$(-1) \approx \$ - 0.28$$

That is, on each \$1 lottery ticket purchased, you should expect to lose about \$0.28. [In general, the question is slightly more complicated than this since it is not reasonable to expect never to need to split the jackpot. Also, in most lotteries of this sort there are other (lesser) prizes that can be won.]

Chapter 16

WALKING

A.   Normal Curves

1. **(a)** $\mu = 83$ lb

    (b) $M = 83$ lb

    (c) $\sigma = 90$ lb. $- 83$ lb. $= 7$ lb

3. **(a)** $Q_1 = 81.2$ lb. $- 0.675 \times 12.4$ lb. ≈ 72.8 lb

    (b) $Q_3 = 81.2$ lb. $+ 0.675 \times 12.4$ lb. ≈ 89.5 lb

5. Since $Q_1 = \mu - 0.675 \times \sigma$,
    $72.8 = \mu - 0.675 \times 12.4$ so that $\mu = 81.17$ lb.
    Thus,
    $Q_3 = 81.17$ lb. $+ 0.675 \times 12.4$ lb. ≈ 89.6 lb.

7. $94.7 = 81.2 + 0.675 \times \sigma$
    $\sigma = 20$ in.

9. **(a)** $\dfrac{72 \text{ in.} + 78 \text{ in.}}{2} = 75$ in.

    (b) $\sigma = 78$ in. $- 75$ in. $= 3$ in.

    (c) $Q_1 = 75$ in. $- 0.675 \times 3$ in. ≈ 73 in.
    $Q_3 = 75$ in. $+ 0.675 \times 3$ in. ≈ 77 in.

11. $\mu \neq M$

13. In a normal distribution the first and third quartiles are the same distance from the mean. In this distribution,
    $\mu - Q_1 = 453 - 343 = 110$ and
    $Q_3 - \mu = 553 - 453 = 100$.

B.   Standardizing Data

15. **(a)** $\dfrac{45 \text{ kg} - 30 \text{ kg}}{15 \text{ kg}} = \dfrac{15 \text{ kg}}{15 \text{ kg}} = 1$

    (b) $\dfrac{54 \text{ kg} - 30 \text{ kg}}{15 \text{ kg}} = \dfrac{24 \text{ kg}}{15 \text{ kg}} = 1.6$

    (c) $\dfrac{0 \text{ kg} - 30 \text{ kg}}{15 \text{ kg}} = \dfrac{-30 \text{ kg}}{15 \text{ kg}} = -2$

    (d) $\dfrac{3 \text{ kg} - 30 \text{ kg}}{15 \text{ kg}} = \dfrac{-27 \text{ kg}}{15 \text{ kg}} = -1.8$

17. **(a)** In a normal distribution, the third quartile is about 0.675 standard deviations above the mean. That is, $Q_3 \approx \mu + (0.675)\sigma$. In this case, it means $278.58 \approx 253.45 + (0.675)\sigma$ so that the standard deviation is $\sigma \approx 37.23$. Hence, the standardized value of 261.71 is
    $\dfrac{261.71 - 253.45}{37.23} \approx 0.22$

    (b) $\dfrac{185.79 - 253.45}{37.23} \approx -1.82$

    (c) $\dfrac{253.45 - 253.45}{37.23} = 0$

19. -0.675

21. **(a)** $\dfrac{x - 183.5}{31.2} = -1$
    $x - 183.5 = -31.2$
    $x = 152.3$ ft

    (b) $\dfrac{x - 183.5}{31.2} = 0.5$
    $x - 183.5 = 15.6$
    $x = 199.1$ ft

(c) $\dfrac{x-183.5}{31.2} = -2.3$

$x - 183.5 = -71.76$

$x = 111.74$ ft

(d) $\dfrac{x-183.5}{31.2} = 0$

$x - 183.5 = 0$

$x = 183.5$ ft

23. $\dfrac{84-50}{\sigma} = 2$

$34 = 2\sigma$

$\sigma = 17$ lb.

25. $\dfrac{50-\mu}{15} = 3$

$50 - \mu = 45$

$\mu = 5$

27. $\dfrac{20-\mu}{\sigma} = -2; \dfrac{100-\mu}{\sigma} = 3$

$20 - \mu = -2\sigma; 100 - \mu = 3\sigma$

$\mu = 20 + 2\sigma$, so $100 - (20 + 2\sigma) = 3\sigma$

$100 - 20 - 2\sigma = 3\sigma$

$80 = 5\sigma$

$\sigma = 16$

$\mu = 20 + 2 \times 16 = 52$

C.   The 68 – 95 – 99.7 Rule

29. *P* is one standard deviation above the mean, and *P′* is one standard deviation below the mean.

$\mu = \dfrac{50+60}{2} = 55$

$\sigma = 60 - 55 = 5$

31. 98.8 is 2 standard deviations above the mean and 85.2 is 2 standard deviations

below the mean.

$\mu = \dfrac{85.2+98.8}{2} = 92$

$\sigma = \dfrac{1}{2}(98.8 - 92) = 3.4$

33. 73.25 is the first quartile and 86.75 is the third quartile.

$\mu = \dfrac{73.25+86.75}{2} = 80$

$Q_3 - \mu \approx 86.75 - 80 = 6.75$

$0.675\sigma \approx 6.75$

$\sigma \approx 10$

35. (a) 95% of the data lies within two standard deviations of the mean. Hence, 2.5% of the data are not within two standard deviations on each side of the mean. So, 97.5% of the data fall below the point two standard deviations above the mean.

(b) $\dfrac{95\% - 68\%}{2} = 13\%$

37. Since 84% of the data lies above one standard deviation of the mean,

$\mu - \sigma = 50.2$ cm.

Thus, $\mu = 6.1$ cm. $+ 50.2$ cm. $= 56.3$ cm.

39. 9.9 has a standardized value of

$\dfrac{9.9-12.6}{4} = -0.675$. Also, 16.6 has a

standardized value of $\dfrac{16.6-12.6}{4} = 1$. So,

25% of the data is below the first quartile of 9.9. Also, 84% of the data is below 16.6. So, 84% - 25% = 59% of the data is between 9.9 and 16.6.

D.   Approximately Normal Data Sets

41. (a) 52 points

(b) 50%

(c) $\dfrac{41-52}{11}=-1, \dfrac{63-52}{11}=1$

68%

(d) $\dfrac{1}{2}(100\%-68\%)=16\%$

43. (a) $Q_1 \approx 52-0.675\times11 \approx 44.6$ points

(b) $Q_3 \approx 52+0.675\times11 \approx 59.4$ points

(c) interquartile range $= Q_3 - Q_1$
$$\approx 59.4-44.6$$
$$= 14.8 \text{ points}$$

45. (a) $\dfrac{99-125}{13}=-2, \dfrac{151-125}{13}=2$

95% have blood pressure between 99 and 151 mm.
$0.95\times2000 = 1900$ patients

(b) 112 is one standard deviation below the mean. 151 is two standard deviations above the mean. The percentage of patients falling between one standard deviation below the mean and two standard deviations above the mean is 68% + 13.5% = 81.5%. 81.5% of the 2000 patients is 1630 patients.

47. (a) $\dfrac{100-125}{13}=-1.92$

approximately the 3rd percentile

(b) $\dfrac{112-125}{13}=-1$

the 16th percentile

(c) $\dfrac{138-125}{13}=1$

the 84th percentile

(d) $\dfrac{164-125}{13}=3$

the 99.85th percentile

49. (a) $\dfrac{11-12}{0.5}=-2, \dfrac{13-12}{0.5}=2$

95%

(b) $\dfrac{12-12}{0.5}=0, \dfrac{13-12}{0.5}=2$

$\dfrac{1}{2}(95\%)=47.5\%$

(c) Because the chance of the bag weighing between 11 and 12 ounces is the same as the chance that it weighs between 12 and 13 ounces, we can use our answer to part (b) and symmetry to obtain an answer of 47.5% + 50% = 97.5%.

51. (a) $\dfrac{11-12}{0.5}=-2$

$\dfrac{1}{2}(100\%-95\%)=2.5\%$

$0.025\times500=12.5 \approx 13$ bags

(b) $\dfrac{11.5-12}{0.5}=-1$

$\dfrac{1}{2}(100\%-68\%)=16\%$

$0.16\times500=80$ bags

(c) 50%
$0.50\times500 = 250$ bags

(d) $\dfrac{12.5-12}{0.5}=1$

$\dfrac{1}{2}(68\%)+50\%=84\%$

$0.84\times500=420$ bags

(e) $\dfrac{13-12}{0.5}=2$

$\dfrac{1}{2}(95\%)+50\%=97.5\%$

$0.975\times500=487.5\approx488$ bags

(f) $\dfrac{13.5-12}{0.5}=3$

$\dfrac{1}{2}(99.7\%)+50\%=99.85\%$

$0.9985\times500=499.25\approx499$ bags

53. (a) $\dfrac{15.25-17.25}{2}=-1$

16th percentile

(b) $\dfrac{21.25-17.25}{2}=2$

97.5th percentile

(c) 75th percentile corresponds to the third quartile.
weight $=17.25+0.675\times2=18.6$ lb

55. (a) $\dfrac{11-8.75}{1.1}\approx2$

97.5th percentile

(b) $\dfrac{12-8.75}{1.1}\approx3$

99.85th percentile

(c) 25th percentile corresponds to the first quartile.
weight $=8.75-0.675\times1.1\approx8$ lb

E.   The Honest and Dishonest Coin Principles

57. (a) $\mu=\dfrac{3600}{2}=1800,\sigma=\dfrac{\sqrt{3600}}{2}=30$

(b) $\dfrac{1770-1800}{30}=-1,\dfrac{1830-1800}{30}=1$
68%

(c) $\dfrac{1}{2}(68\%)=34\%$

(d) $\dfrac{1860-1800}{30}=2$

$\dfrac{1}{2}(95\%)=47.5\%$

$47.5\%-34\%=13.5\%$

59. $\mu=\dfrac{7056}{2}=3528,\sigma=\dfrac{\sqrt{7056}}{2}=42$

(a) $\dfrac{3486-3528}{42}=-1,\dfrac{3570-3528}{42}=1$
68%

(b) $\dfrac{1}{2}(100\%-68\%)=16\%$

(c) $\dfrac{1}{2}(68\%)+50\%=84\%$

61. (a) $\mu=600\times0.4=240,$
$\sigma=\sqrt{600\times0.4\times(1-0.4)}=12$

(b) $Q_1\approx240-0.675\times12\approx232$
$Q_3\approx240+0.675\times12\approx248$

(c) $\dfrac{216-240}{12}=-2,\dfrac{264-240}{12}=2$
0.95

63. (a) $p=\dfrac{1}{6}$

$\mu=180\times\dfrac{1}{6}=30,\sigma=\sqrt{180\times\dfrac{1}{6}\times\left(1-\dfrac{1}{6}\right)}=5$

(b) $\dfrac{40-30}{5}=2$

$\dfrac{1}{2}(1-0.95)=0.025$

(c) $\dfrac{35-30}{5}=1$

$\dfrac{1}{2}(0.68)=0.34$

65. We assume that the defects are distributed normally so that the mean number of defects on a given day is $\mu=90,000\times0.10=9,000$ by the Dishonest-Coin principle. The standard deviation is given by $\sigma=\sqrt{90,000\times0.10\times(1-0.10)}=90$. So, 9,180 is 2 standard deviations above the mean. It follows that the machine needs to be recalibrated 2.5% of the time.

JOGGING

67. **(a)** weight $=17.25+1.65\times2=20.55\,\text{lb}$

   (b) weight $=17.25-0.25\times2=16.75\,\text{lb}$

69. **(a)** $\dfrac{17.75-17.25}{2}=0.25$

   60th percentile

   (b) $\dfrac{16.2-17.25}{2}=-.52$

   30th percentile

71. **(a)** score $=520+(0.675)(115)=597.625\approx600$ points

   (b) $\dfrac{750-520}{115}=2$. Thus, a score of 750 is 2 standard deviations above the mean. This means a score of 750 is at the 97.5^{th} percentile.

73. **(a)** The 90th percentile of the data is located at $\mu+1.28\times\sigma=65.2+1.28\times10=78$ points.

   (b) The 70th percentile of the data is located at $\mu+0.52\times\sigma=65.2+0.52\times10=70.4$ points.

   (c) The 30th percentile of the data is located at $\mu-0.52\times\sigma=65.2-0.52\times10=60$ points.

   (d) The 5th percentile of the data is located at $\mu-1.65\times\sigma=65.2-1.65\times10=48.7$ points.

75. $\mu=\dfrac{n}{2}$

   There is a 95% chance when X is within 2 standard deviations of the mean.

   $15=2\sigma$

   $\sigma=7.5$

   $7.5=\dfrac{\sqrt{n}}{2}\ \left(\text{using }\sigma=\dfrac{\sqrt{n}}{2}\right)$

   $15=\sqrt{n}$

   $n=225$

77. For $p=\dfrac{1}{2}$, $\mu=np=\dfrac{n}{2}$, and

   $$\sigma=\sqrt{np(1-p)}=\sqrt{n\cdot\dfrac{1}{2}\cdot\dfrac{1}{2}}=\sqrt{\dfrac{n}{4}}=\dfrac{\sqrt{n}}{2}.$$